War Planning 19

The major European powers drafted war plans before 1914 and executed them in August 1914; none brought the expected victory by Christmas. Why? This tightly focused collection of essays by international experts in military history reassesses the war plans of 1914 in a broad diplomatic, military, and political setting for the first time in three decades. The book analyzes the war plans of Austria-Hungary, France, Germany, Great Britain, Italy, and Russia on the basis of the latest research and explores their demise in the opening months of World War I. Collectively and comparatively, these essays place contingency war planning before 1914 in the different contexts and challenges each state faced as well as into a broad European paradigm. This is the first such undertaking since Paul Kennedy's groundbreaking *War Plans of the Great Powers* (1979), and the end result is breathtaking in both scope and depth of analysis.

Richard F. Hamilton is Professor Emeritus at The Ohio State University. He is the author of *President McKinley and the Coming of War, 1898*; *President McKinley and America's "New Empire"*; and *The Origins of World War I.*

Holger H. Herwig is Canada Research Chair in Military and Strategic Studies at the University of Calgary. He is the author of *The First World War: Germany and Austria-Hungary 1914–1918*; *War Memory and Popular Culture*; *The Origins of World War I*; and *The Marne: 1914*.

39 £2.49 C10 Sc

War Planning 1914

Edited by

RICHARD F. HAMILTON

Ohio State University

HOLGER H. HERWIG

University of Calgary

CAMBRIDGE
UNIVERSITY PRESS

32 Avenue of the Americas, New York NY 10013-2473, USA

Cambridge University Press is part of the University of Cambridge.

It furthers the University's mission by disseminating knowledge in the pursuit of education, learning and research at the highest international levels of excellence.

www.cambridge.org
Information on this title: www.cambridge.org/9781107635128

First published 2010
First paperback edition 2013

A catalogue record for this publication is available from the British Library

Library of Congress Cataloguing in Publication data

War planning 1914 / edited by Richard F. Hamilton, Holger H. Herwig. – 1st ed.
p. cm.
Includes bibliographical references and index.
ISBN 978-0-521-11096-9 (hbk.)
1. Military planning – Europe – History – 20th century. 2. World War, 1914–1918 – Political aspects. 3. World War, 1914–1918 – Causes. 4. Strategy – History – 20th century. 5. Europe – Politics and government – 1871–1918 – Decision making.
I. Hamilton, Richard F. II. Herwig, Holger H. III. Title.
U155.E85W37 2009
355′.0335409041–dc22 2009021686

ISBN 978-0-521-11096-9 Hardback
ISBN 978-1-107-63512-8 Paperback

Contents

List of Maps

Contributors

Robert A. Doughty
Department of History
U.S. Military Academy
West Point, New York

John Gooch
School of History
University of Leeds
Leeds, United Kingdom

Richard F. Hamilton
The Mershon Center
The Ohio State University
Columbus, Ohio

Holger H. Herwig
Centre for Military and Strategic Studies
University of Calgary
Calgary, Alberta, Canada

Günther Kronenbitter
Department of History
Emory University
Atlanta, Georgia

Bruce W. Menning
Department of Joint and Multinational Operations
U.S. Army Command and General Staff College
Ft. Leavenworth, Kansas

Annika Mombauer
History Department
The Open University
Milton Keynes, United Kingdom

Keith Neilson
Royal Military College of Canada
Kingston, Ontario, Canada

Acknowledgments

This book began with a brainstorming conference at The Ohio State University in 2005. We are deeply indebted to Richard Herrmann, Director of the Mershon Center for International Security Studies, for making that possible, and to the Center's able staff, most especially to Beth Russell, for their continued generous support of the project.

The Austro-Hungarians and the Germans annually updated their war plans for about a decade; France went through seventeen and Russia through nineteen major iterations of their war plans before 1914. Thus, not surprisingly, we also experienced changes and alterations in our own "war planning," most notably through academic moves, child birth, family illness, and the vicissitudes of archival availability. We are especially grateful to Annika Mombauer for coming to our rescue in the chapter dealing with the German Plans for War and to Bruce Menning for kindly sharing some of his unique war plans maps with us. We thank all the contributors for their patience in graciously replying to our editorial queries and suggestions, and especially the original conference participants for kindly updating their bibliographies and notes.

Richard F. Hamilton
Holger H. Herwig

1

War Planning: Obvious Needs, Not So Obvious Solutions

Richard F. Hamilton

Wars are curious, puzzling, problematic events. Sensible people everywhere know that wars can be extremely costly in lives, property, and money. And they know that their outcomes are uncertain. Yet they still happen. Few other human activities have stimulated as much thought as the simple question: Why do wars happen?[1] More often than not, the concern is with specific wars such as World War I, once called the Great War. The basic question: what happened in August 1914? Or, more precisely, why did the leaders of Europe's major powers choose war?

A formal answer to that question, something discussed in a previous work, *The Origins of World War I*, is that in each case a small coterie, the nation's leaders, made the decisions that took the nation to war. Those leadership groups assessed current situations, defined the threats, considered alternatives, and chose war as their most appropriate option, either initiating action or responding to another nation's initiative. In each case, moving a step beyond that formal statement, the participants had specific strategic agendas. Austria-Hungary's leaders were seeking to prevent dissolution of the empire, a threat they sensed as stemming from Serbia's initiatives. Germany's leaders saw the threat of a two-front war with France and Russia arming and becoming ever more dangerous. For Germany's leaders the events of July and August 1914 provided a twofold opportunity, to protect their most valued ally, Austria-Hungary, and to remove the threat posed by the *entente* powers. In summary, one German leader declared that it was "now or never." Russia's leaders saw their nation threatened, primarily, by the central European allies,

[1] For recent overviews, see Appendix, "On Wars: General."

Germany and Austria-Hungary. Our view is that the decisions made by those coteries, based on their readings of threats and judgments of resultant needs, brought about the Great War. Our argument, in short, focuses on decision-making groups and on what we termed the strategic causes.[2]

Following the basic decision, saying "yes" to the option of war, the immediate concern for any decision-making coterie would be that of the appropriate war plan: how should the war be conducted? Most modern (or developed) nations would have several contingency war plans ready at hand. In this scenario, one would expect to find a division of labor and some necessary coordination. A nation's top leaders, monarchs, presidents, and ministers, would ordinarily make the initial decision, the choice of war. War plans would, presumably, have been developed by military leaders, the trained specialists. Leaders, civilian and military, would then coordinate the first steps of implementation, justifying the action, mobilizing the needed forces, transforming the economy, and so on.

The preceding paragraph contains several plausible inferences, some easy, logical conclusions. But such assumptions, beginning with the term "war plan," are seriously misleading. The term itself, war plan, sounds both easy and obvious. But such plans are extraordinarily complicated. They involve the movements and aims of armies, of navies, and, as of 1914, of rudimentary air forces. A plan must deal with the movement of large forces (horse-drawn wagons, motor vehicles, trains, and ships). It must deal with the supplying of those same units. One would have to organize intelligence (i.e., spy), assess the findings, and adjust plans accordingly. A naval plan must consider confronting the enemy (or enemies). It must also arrange for a blockade to prevent the passage of supplies to the enemy and, simultaneously, must prevent the enemy from blocking one's own supply routes. With large units in movement, some in enemy territory, one must consider and coordinate the means of communication.

The term "plan," a static concept suggesting fixity, is inappropriate. In a war, especially in the opening phases, to quote Heraclitus, "All is flux, nothing is stationary." The appropriate term, accordingly, is planning whereby one is dealing with an ongoing process. Still another cautionary note: given the complexities, close coordination of these diverse activities

[2] Richard F. Hamilton and Holger H. Herwig, eds., *The Origins of World War I* (Cambridge, 2003).

would seem a strong imperative. But, as will be seen, the requirement of secrecy, to ensure that none of this crucial information reaches the enemy, limits such efforts.

Five of the six major European powers began the Great War following the principal directions of their respective war plans. Many educated persons are likely to know "the basics" of one such plan, Germany's Schlieffen Plan. Apart from the handful of specialists, however, few would know anything of the comparable Austro-Hungarian plan, the first to be implemented. And few would know anything of the Russian, French, or British plans, or that of the exceptional case, Italy. Apart from the work of a few specialists, war plans and planning processes have not been of great interest to those concerned with World War I.[3]

Some words of caution should be noted. Those clear and unambiguous expressions – "war plans" and "the Schlieffen Plan" – can be seriously misleading. They suggest the existence of definitive statements, ones that have been tested, reviewed, and agreed upon. All that remains, presumably, is implementation. When the need arises, the nation's decision-makers reach for "the plan" and, following "its" prescriptions, proceed according to the directions contained there. Depictions of final definitive plans or of leaders following "its" prescriptions, as will be seen, could hardly be more mistaken.

Instead of a static image – "the plan" – it is best to think in terms of a process, that is, of a continuous, ever-changing planning effort. The French experience illustrates the point – their option in August 1914 was entitled Plan XVII. The Russians proceeded with a modified version of their Mobilization Schedule No. 19A. The processes, moreover, are often disorderly with conflicting perceptions of threat and serious differences with regard to appropriate responses. Some accounts suggest orderly planning processes directed by rational calculating decision-makers. But, as will be seen, much of the planning is better described as murky with poorly informed and anxious participants making last-minute reversals, in some cases extemporizing major changes. The title here – *War Planning* – is intended to signal this fact, a continuous process, as opposed to the notion of some final, definitive, and binding plan. The word plan appears frequently in the following pages but, keeping in mind the

[3] Two notable exceptions: Paul M. Kennedy, ed., *The War Plans of the Great Powers, 1880–1914* (London and Boston, 1979); and John H. Maurer, *The Outbreak of the First World War: Strategic Planning, Crisis Decision Making, and Deterrence Failure* (Westport, CT, 1995).

just-reviewed difficulties, readers should remember the tentative or provisional character of many such accomplishments.

War plans, it should be noted, are a relatively recent phenomenon, something that came late in the late nineteenth century. Alexander probably had no such comprehensive plan when he set out from Macedonia to conquer the world. Charles XII, the king of Sweden, probably had no such plan when, in 1708, he decided to invade Russia. When first setting out, he had probably never heard of Poltava, the scene of the decisive battle. Until well into the nineteenth century, military commanders moved their troops about the landscape, either pursuing or avoiding, during which time they studied their opportunities, the task being to discover the best place and time to engage the enemy.

No specific "war plans" have been found for the campaigns of George Washington during the American Revolutionary War. His was a system of *ad hoc* reactions in the face of a superior British force of 32,000 men, 130 warships, and more than 300 cannon. Washington, who was described by Ambrose Serle, General William Howe's top aide, as the "little paltry colonel of a militia of bandits," remained on the defensive in the last months of 1776.[4] Defeats at Brooklyn Heights, Harlem Heights, White Plains, and Fort Washington forced the American commander to avoid further engagement. He and his men crossed the Hudson, marched across New Jersey, crossed the Delaware and halted, momentarily, in Pennsylvania. Then, sensing an appropriate moment, Washington and his forces re-crossed the Delaware to surprise and defeat the Hessians in Trenton. Several days later, in another surprise move, he defeated a small British force at Princeton. Then, enduring incredible winter hardships at his base in Morristown early in 1777, Washington engaged in guerilla tactics, hitting the British and their supply lines as opportunity allowed. Only the next year, with a revived Continental Army, was he able to engage the British in head-on conventional European warfare and to defeat them – which he did at Monmouth, New Jersey.

Napoleon Bonaparte did not have a "war plan" when he set out to conquer the rest of Europe. It was not part of his plan that he would bring his forces specifically to Austerlitz or Jena. Later in the century, when the size of armies had increased enormously, when the weight and volume

[4] Bruce Chadwick, *George Washington's War: The Forging of a Revolutionary Leader and the American Presidency* (Naperville, Ill., 2004), the quotation, p. 4. See also David Hackett Fischer, *Washington's Crossing* (New York, 2004); and David McCullough, *1776* (New York, 2005).

of the weaponry increased with accompanying logistical needs, extensive and detailed planning became necessary.[5]

Most accounts of World War I follow a familiar narrative. After reviews of the background events and of the July crisis, they turn to the initial operations of the war, beginning with Austria-Hungary's move into Serbia. Then much attention is given to Germany's sweep through Luxembourg and Belgium and the move into France, following a modified Schlieffen Plan. That effort ended in failure with Germany's defeat in the First Battle of the Marne, a failure that introduced (or "set the stage for") consideration of the subsequent four years of the Great War. Within weeks, it was evident that *all* of the war plans of the major powers were seriously flawed, some in astonishing ways.

The war began with the Austro-Hungarian move. The plan was similar to the Schlieffen Plan, it too assuming a two-front war with a struggle first against a weak enemy, Serbia, and then one against a much stronger enemy, Russia. The armed forces would be moved first for the quick and total defeat of Serbia. They would then be shifted to Galicia in the northeast to confront the Russians, who, being slower to mobilize, would now be coming to the aid of their Slavic ally. Fully aware of the likely Russian response, Austria-Hungary's first requirement, one of utmost importance, was to secure a promise of support from their German ally. After securing that assurance, the Dual Monarchy's forces then invaded Serbia moving from Bosnia. The Russian move into Galicia came sooner than expected. After "frantic appeals" from the German chief of staff, Austria-Hungary broke off the Serbian campaign, leaving that enemy undefeated and moved its armies to Galicia. There, a hopelessly mismanaged offensive brought a catastrophic defeat to the Austro-Hungarian army and the loss of a third of its forces.

Russia also faced a two-front problem. With two enemies in the West, Germany and Austria-Hungary, an appropriate plan was developed. Russia's first move in the war was against Austria-Hungary with the invasion of Galicia. Then its troops moved against Germany, entering East Prussia *en masse*, dividing forces as they passed through the Masurian lakes region. One aspect of this operation brought much deprecating comment, both then and later: the Russians were communicating "in the clear." Messages sent to help coordinate movement through the lake district were not encoded, a "planning failure" that helped to produce an important German victory in the Battle of Tannenberg. But the problem

[5] A brief review of this transformation of military procedures appears here in Chapter 8.

was general – "in France the German Army did precisely the same, with identical results."[6]

For some years, Germany's leaders had anticipated a two-front war, their enemies being France in the West and Russia in the East. The Schlieffen Plan had been designed to address that problem. Like the Austrian and Russian designs, this too called for "one-two punches," first a quick move with overwhelming forces against weaker France followed by the transfer of forces for the longer engagement on the eastern front. The move into France proceeded with initial success, all apparently proceeding according to schedule. But then, making the great sweep to the north of Paris for the attack on the rear of the French forces in Lorraine and Alsace, French leaders saw their opportunity. The Germans were "giving them their flank." The French mobilized all available forces, attacked the exposed German right wing, and won that famous first battle. The Germans were then forced to retreat and to dig in. The long four-year battle of the trenches now began. The plan was brilliant, but seriously flawed. Also, a second step of planning was missing. What should be done if the first step failed?

French planning is perhaps the most problematic of the six to be discussed. French leaders knew the basics of the Schlieffen Plan, namely, that the Germans would invade through Belgium. Yet they placed the bulk of their forces to the East, facing Lorraine and the Ardennes, planning to begin with offensive operations there. Here too there was a serious planning error. French artillery was "state of the art," a fast-firing, 75-mm weapon. But it had a relatively short range, most notably a shorter range than German heavy artillery. It also had a flat trajectory and Lorraine was hilly. To reach an enemy on the other side of those hills, they needed weapons with high, arching fire. A more serious problem, one that would have pernicious effects, was the acceptance of the "offensive doctrine."

The British "plan" called for a quick movement of units across the Channel, to be placed on the western edge of the allied front. The Germans moved with massive numbers, which overwhelmed the modest British forces. On the first day of their engagement, the Empire lost more soldiers than had been lost in combat on any day in the previous century.

Italy was allied with Germany and Austria-Hungary. Its leaders had promised Germany that they would send an army to the Rhine in the

[6] For a detailed account of the problems involved, see John Ferris, ed., *The British Army and Signals Intelligence during the First World War* (Phoenix Mill, UK, 1992), Introduction.

event of war. The presence of those numbers facing the French would have freed up more German units for the sweep through Belgium and, perhaps, would have brought the anticipated "decisive" victory. But in Germany's "hour of need," Italy's leaders instead declared that nation's neutrality.

Italy's and Germany's leaders had also given some serious last-minute consideration to two other plans, an attack on France through the Alps and one in Provence on the Mediterranean coast. Given the secrecy and the death of a key decision-maker in the midst of the crisis, at one point, two groups of Italy's leaders were moving to implement different plans. The decision to avoid immediate engagement, moreover, hid a serious flaw in all of these options. Ninety percent of Italy's coal supply came from Great Britain. The most likely alternative source would have been Germany. But no consideration had been given to the availability of high-grade coal, of the need for locomotives and coal cars, of the schedules, or of the routes to be taken.

The development of those war plans and of the attendant pathologies are the subject of this book. The aim is to examine the planning processes and the resulting plans of the six major European powers.

The most appropriate theoretical framework to deal with such decision-making is the elite theory (or, for short, elitism).[7] It recognizes

[7] For an overview, see G. Lowell Field and John Higley, *Elitism* (London, 1980); also, Mattei Dogan and John Higley, eds., *Elites, Crises, and the Origins of Regimes* (Lanham, Md., 1998).

Modern formulations of elite theory appeared first, more than a century ago, in the works of Gaetano Mosca, Vilfredo Pareto, and Robert Michels. Their writings were translated and published in English but never gained much attention in the English-speaking world. Possibly the most famous English-language work in this tradition is by C. Wright Mills, *The Power Elite* (New York, 1956). It contains hundreds of pages about elites in American society, principally those of business and the military, but the work provides no serious evidence about their power or their relationships, the basic problem being one of assertion without evidence. The work remained in print for more than four decades, the last edition appearing in 2000. Suzanne Keller wrote on the autonomy of different elite groups and the problems of their coordination, in *Beyond the Ruling Class: Strategic Elites in Modern Society* (New York, 1963).

Elitism has never gained much attention in the social sciences (apart from the Mills volume) or in history. Field and Higley point up the problem in their opening chapter, one entitled "Elitism in eclipse." For later overviews on the theory and its uses, see Eva Etzioni-Halevy, *The Elite Connection: Problems and Potential of Western Democracy* (Cambridge, UK, 1993); also two works edited by Etzioni-Halevy, *Classes and Elites in Democracy and Democratization* (New York, 1997); and a special issue of the *International Review of Sociology*, 9,2 (1999).

that complex, economically advanced societies develop separate specialized organizations to accomplish necessary or desired goals. Those organizations, sometimes referred to as institutions, include political, economic, and military agencies, as well as those dealing with education, mass communication, religious matters, and so forth. Within each of those categories one would ordinarily find further specialties such as army, navy, and later, air force, as well as various intelligence branches. Those agencies developed somewhat differently in each nation. Also, the pacing of the achievements varied, some being early, some late developments.

Each of these institutions would have a leadership group, the elite, which is central to the theory. Those persons, typically, would have received some initial training in specialized professional schools. Later they would have undergone extensive on-the-job training, moving up the organizational hierarchy to positions that required greater knowledge, oversight, and responsibility.

Given the diverse career lines, each with distinctive training and experience, it follows that the various elites would have different concerns and priorities. Each elite, moreover, would have some degree of autonomy, that is to say, some power or resources allowing the defense or furtherance of its perceived interests. Over time, some understandings, formal or informal, would be developed defining the proper or appropriate tasks of each institution. One such experience would be the European secularization conflict, the centuries-long struggle over church-state relations that ultimately removed religious elites from most tasks of governance.

Given their separate training, careers, and locations it follows that the various elites would not ordinarily have regular contact. And it follows further that some efforts of coordination by these elites, either regular or occasional, would be necessary. Although the logic is easy, such statements are best viewed as hypotheses. Investigation – a testing of those hypotheses – is always appropriate. Did coordination actually occur? And if it did, what was the result?

It is easy also to assume the dominance of a given elite. Economic elites, the owners and managers of leading industrial firms and of the major banks, are often assumed to be all-powerful in modern capitalist societies. The "force" of a logical argument, unfortunately, is often so strong that investigation is discouraged or neglected entirely. The "obvious" (or "compelling") logic allows one to by-pass the onerous tasks of research. The founders of a major social theory, for example, one first expounded in 1848, dealt with the complexities of business-government relations with a simple declaration: "The executive of the modern state is but a committee for managing the common affairs of the whole bourgeoisie."[8]

In their extensive later writings, Karl Marx and Friedrich Engels neglected the research task entirely, never showing how "the whole bourgeoisie" selected that "executive committee." It was never demonstrated that "they" were directing the governments of Lord Palmerston, Benjamin Disraeli, or William Gladstone. In 1854, Marx and Engels could have established the point by presenting evidence showing the "whole bourgeoisie" of Britain and France directing their respective governments to intervene in the Crimea to fight a very costly war. But they did not undertake that task, the declaration of the obvious logic being considered fully sufficient. The logic of the argument, put simply, is that wars in the modern era stemmed from the demands of the capitalists, demands that somehow "reflected" (or "expressed") the needs of the capitalist system.[9]

During the Great War, a revision of the Marxist position was published by Vladimir Ilych Lenin in his *Imperialism: The Highest Stage of Capitalism*. He argued that capitalism in its latest form, *Finanzkapital*, must seek out new settings for investment, hence the need for colonies with direct rule as the most "convenient form." To achieve those aims, the major capitalist nations chose war; for them, it was a necessity, a life-or-death matter. The World War, "on the part of both sides," Lenin wrote, was the "proven" result. This work later found an important sponsor, the Soviet Union, which subsidized translations, publication, and gave it worldwide dissemination. It would be "read and studied by millions of people who know no other book on imperialism."[10] This work too was based on a plausible logic as opposed to appropriate research and evidence.

[8] Karl Marx and Frederick Engels, *The Communist Manifesto*, in the *Collected Works* (New York, 1976), vol. 6, pp. 477–519, quotation, p. 486.

[9] Marx was living in London at this time writing weekly articles for Horace Greeley's New York *Tribune*. One of these gives a strikingly different explanation, declaring that the Crimean crisis was produced by the "conflict between the Latin and Greek churches" in the Near East and that France's intervention was the work of Emperor Louis Napoleon who was seeking Catholic support. Marx's summary, all italicized, reads: "The Bonapartist usurpation, therefore, is the true origin of the present Eastern complication." From his "Russian Diplomacy . . . ," *Tribune*, 27 February 1854, in Marx and Engels, *Collected Works*, Vol. 12, pp. 615–616.

[10] V. I. Lenin, *Collected Works* (46 vols., Moscow, 1960), vol. 22, pp. 185–304. The citation is from Norman Etherington, *Theories of Imperialism: War, Conquest and Capital* (London, 1984), p. 160. For explication and assessment, see Richard F. Hamilton, *Marxism, Revisionism, and Leninism* (Westport, Conn., 2000), Ch. 4.

The Marxist theory challenged a basic claim of the liberal position. An impressive array of liberal thinkers had pointed to the growing internationalism of trade and the resulting interdependencies. Those facts, it was argued, provided powerful incentives for businessmen to oppose war. Among those arguing this position were Baron de Montesquieu, Adam Smith, Immanuel Kant, Thomas Paine, David Ricardo, Jeremy Bentham, James Mill, John Stuart Mill, Herbert Spencer, Richard Cobden, John Bright, J. A. Hobson, and, last but not least, a recipient of the Nobel Peace Prize, Norman Angell.[11]

John Stuart Mill, writing in 1848, declared that commerce "is rapidly rendering war obsolete, by strengthening and multiplying the personal interests which act in natural opposition to it." The liberal position held that it was the old regimes, those led by monarchs and aristocracies, that chose to make war. The new regimes, the republican arrangements, responding to significant and growing business influence, would bring an end to wars. Like Marxism, this position was also based on an "obvious" logic and again little investigation was undertaken to establish the basic claim. One similarity ought to be noted: both liberalism and Marxism assumed the dominant influence of business elites. In both theories, "they" were directing the efforts of the political leaders. In both cases, the conclusion was based on a "compelling" logic, which is to say it was presented without benefit of serious research.

The Great War presented an overt challenge to the widely held optimistic liberal assumptions. That event thoroughly discredited the central claim of the liberal position, that international trade and interdependence meant peace. That position, perhaps not surprisingly, would now be treated with disdain, as a curious naïve illusion that no intelligent person could possibly accept.

We have an unexpected convergence here. After 1914–1918, both Marxism and liberalism (the new revised version thereof) were in substantial agreement that "big business" was the dominant elite in advanced capitalist regimes. And both agreed also that "big business" somehow accepted or favored war as an instrument of policy. Because business was "in power," the existence of wars must mean, as expressed in three familiar clichés, that "they" were operating behind the scenes, pulling the strings, and calling the shots. "They" must have pushed for those interventions. The frequent resort to metaphor indicates a serious problem

[11] See Michael E. Howard, *War and the Liberal Conscience* (London, 1978). The Mill quotation in the following paragraph is from p. 37.

with such allegations, namely, a lack of relevant evidence. If the persons making the argument had evidence, they would have presented it rather than resorting to literary images, with the politicians as puppets. The use of analogy and metaphors, it should be noted, changes the subject thus avoiding the topic under consideration.

In *The Origins of World War I* we showed, among other things, that in at least three of the major participating nations – Austria-Hungary, Germany, and Great Britain – leaders of "big business" were adamantly opposed to the prospect of war. The correlated lesson, clearly, is that the political elites were not bound by those objections; they overruled the business leaders. Those findings indicate that the political leaders were autonomous or, put differently, that they were "in power." And the correlated lesson, obviously, is that at least in this contention "big business" proved to be powerless. Those findings mean that differentiated conclusions were, and are, appropriate. The liberals were right about the attitudes of business leaders. But, like the Marxists, they were mistaken in their conclusion that business leaders were "in power."[12]

One might expect that political and military elites would collaborate in the development of war plans, that too being an "obvious" logical inference. But again, against the easy inference we have some contrary evidence. First, two specialist historians report that in the United States "during the years from 1898 to 1917 the [naval] strategists were never once informed of the major objectives of the Administration's diplomacy. In the absence of such guidance, the strategists simply had to establish the basic premises of policy for themselves."[13]

United States army strategists faced the same problem, the absence of direction by political elites and as a result had to infer those likely basic premises. One account tells of the result:

> Until February 1917 no thought whatever had been given to co-operating with the allies on land or on the high seas. Not even a rough plan existed to provide for the eventuality of sending an American expeditionary force to Europe. The war plans held in readiness by the army included an American invasion of Canada (1912–13)

[12] Hamilton and Herwig, *Origins*, pp. 503 ff. No sufficient evidence was found to allow conclusions about business preferences in the other involved nations.

In 1898, business leaders in the United States opposed the nation's intervention to aid Cuban insurgents in what was to be the Spanish-American War. But there too, the political leaders decided otherwise, see Richard F. Hamilton, *President McKinley and the Coming of War, 1898* (Piscataway, NJ, 2006), Ch. 4.

[13] John A. S. Grenville and George Berkeley Young, *Politics, Strategy and American Foreign Policy* (New Haven, Conn., 1966), p. 294.

and also envisaged such possibilities as an attack on New York by Great Britain (March 1915) and the defence of the Pacific coast from a Japanese invasion (February 1915–March 1917). The first plans for an American expeditionary force to Europe, not drawn up until February 1917, were based on the possibility of invading Bulgaria through Greece, and of invading France in the rear of the German armies in alliance with Holland. All these plans were only fit for the waste-paper baskets of the War Department. Consequently the full impact of American intervention was delayed for many months. Wilson had not provided the leadership to prepare the nation effectively for a war which until the very last he regarded as disastrous, while the strategists had failed to consider eventualities which their president virtually refused to envisage.[14]

Another instance where that "obvious" coordination was lacking comes from the British experience. In the weeks prior to the outbreak of the Great War, the most urgent cabinet discussions occurred between 23 July and 4 August 1914. During most of that period the leaders of the General Staff were not informed of the Cabinet's deliberations. On 2 August Field Marshal Sir John French telephoned Lord Riddell of the Newspaper Proprietors' Association and asked: "Can you tell me, old chap, whether we are going to be in this war? If so, are we going to put an army on the Continent and, if we are, who is going to command it?"[15]

Italy provides another instance in which "obvious" coordination was lacking. The chief of staff, General Alberto Pollio, suffered a heart attack on 28 June 1914, the day Archduke Franz Ferdinand was assassinated. Misdiagnosed and mishandled, Pollio died on 1 July. There was dissention over the choice of his successor with the result that General Luigi Cadorna, who was first appointed on 20 July, did not take office until 27 July. Totally unaware of the latest cabinet plans, following a previous war plan, he gave orders for the mobilization of units in the Mediterranean Alps, this for a planned action against France. He also brought the Third Army corps up to full strength; it was to be moved to Alsace and Lorraine to support Germany, Italy's ally. But the political leaders, as indicated, had something else in mind. On 31 July, at a very late point in the crisis, the Italian cabinet reviewed matters and decided that "neutrality was the only possible policy." Two days later, the king returned to Rome and approved that declaration. The decision was made public on 3 August, the day German forces crossed into Belgium.[16]

[14] John A. S. Grenville, "Diplomacy and War Plans in the United States, 1890–1917," Ch. 1, pp. 36–37, of Kennedy, ed., *War Plans of the Great Powers*.

[15] Lord Riddell, *Lord Riddell's War Diary* (London, 1933), p. 6.

[16] Hamilton and Herwig, *Origins*, pp. 361–2, 369, 374.

This brief review yields another obvious but often neglected lesson: at the beginning of any investigation one should consider a range of possibilities, a range of alternative hypotheses, some of them not immediately obvious. Put differently, one should consider other possible logics, other rationales. And when possible, those alternatives should be "put to the test."[17]

Why would leaders *not* coordinate? One consideration, possibly the most important justification, is the need for secrecy. The more persons involved, the greater the likelihood of plans being revealed. We have an example from the July crisis: Austria-Hungary and Germany were putting their plans into operation, proceeding with great secrecy and cunning so as to enhance their chances of success. At one point, the chatty German ambassador, Hans von Flotow, told Italy's foreign minister, Antonio di San Giuliano, that Vienna was about to use force against Serbia. The next day, San Giuliano learned further details about Austria-Hungary's plan, including the support of Germany. This information was then transmitted elsewhere, most importantly to Italy's ambassador in St. Petersburg. The message was intercepted and decoded and became available to the Russian and French leaders who were then conferring in the Russian capital. This information presumably allowed the two *entente* powers time for some additional last-minute planning.[18]

In the years before to the Great War, the French Foreign Ministry "refused to share diplomatic plans with the War Ministry." The secrecy was compounded in that the War Ministry "refused to communicate information to politicians it considered incompetent."[19] In 1902, the French Foreign Ministry signed a convention with Italy that provided some assurance of peace on their shared frontier. For reasons of security and interdepartmental rivalry, however, this was revealed to no one outside of a narrow circle of civilian leaders. France's military leaders first learned of the agreement some seven years later and, understandably, were furious. They had stationed two corps on the Italian frontier to guard against a nonexistent threat. The War Ministry, in turn, had its own

[17] See T. C. Chamberlin, "The Method of Multiple Working Hypotheses," *Science* 148 (1965): 754–9.

[18] Hamilton and Herwig, *Origins*, p. 372. Also, R. J. B. Bosworth, *Italy, the Least of the Great Powers: Italian Foreign Policy before the First World War* (London, 1979), p. 386; and William A. Renzi, *In the Shadow of the Sword: Italy's Neutrality and Entrance Into the Great War, 1914–1915* (New York, 1987), pp. 61–2.

[19] Hamilton and Herwig, *Origins*, pp. 260–1; M. B. Hayne, *The French Foreign Office and the Origins of the First World War, 1898–1914* (Oxford, 1993), p. 35.

secrets. It held regular talks with Britain's military leaders in this period. Acting out of the same concern, secrecy, they did not reveal the details to the Foreign Ministry although they did provide the "big picture."

This suspicion and hostility between the two key government agencies led to a sharp confrontation in 1911. The military leader, General Joseph Joffre, "believed that the army ought to be briefed on foreign policy *before* drafting its war plans." But Premier Joseph Caillaux was "unwilling" to share his government's views. The matter was the subject of a heated discussion in October that year. Joffre and Alexandre Millerand, the war minister, were arguing with Foreign Minister Justin de Selves over whether the army had a right to information from the Quai d'Orsay. The president, Armand Fallières, joined in and supported Joffre's request for prior diplomatic guidance. At that point, Premier Caillaux intervened and told the president to "shut up."[20]

The French Foreign Ministry at one point had succeeded in breaking Germany's code and thus was privy to much of their likely enemy's plans and movements. In a careless moment, however, Premier Caillaux boasted to the Germans of this cryptographic success. His flip remark brought the loss of France's "single most valuable source of foreign intelligence."[21] The loss was not permanent, since the French broke other codes. France's problem in August 1914 was not lack of intelligence; it was Joffre's refusal to believe the intelligence reports.

These single-paragraph summaries are intended to show the difficulties involved in coordinating planning efforts. Each of the instances reviewed here, understandably, is much more complicated than can be indicated in a brief summary. The French leaders were enormously relieved when the Italians opted out of their commitments to Germany but they could not be sure what Italy's leaders would do in an actual engagement. Removal of troops from the Italian frontier, moreover, would have given away the secret, making clear that some new arrangement had been made. The purpose of these reviews, it should also be noted, is analytical, not judgmental. The aim is to point out the complexities, the difficulties hindering exemplary decision-making.

Government turnovers pose serious difficulties for the maintenance of secrecy. The replacement of a government might remove a dozen or so

[20] Hamilton and Herwig, *Origins*, pp. 260–1; Samuel R. Williamson, Jr., *The Politics of Grand Strategy: Britain and France Prepare for War, 1904–1914* (Cambridge, Mass., 1969), p. 211.

[21] Christopher Andrew, "France and the German Menace," in Ernest R. May, ed., *Knowing One's Enemies: Intelligence Assessment before the Two World Wars* (Princeton, 1984), pp. 130–1.

"knowledgeable" politicians from office and bring in a similar number of newcomers who, somehow or other, ought to be provided with at least some "basic" information. The turnover problem was (and is) much more serious in parliamentary regimes. As compared to the high continuity seen in the authoritarian regimes, republican governments typically experienced high turnover of both civilian and military leaders.[22]

Those familiar "obvious" assumptions are challenged by still another consideration, by a simple fact of life. With only a small number of persons knowing the details of a war plan, the death of any one of them might have unforeseeable disruptive consequences as just seen in the case of General Pollio. Similarly, the death of Foreign Minister di San Giuliano at Rome on 16 October 1914 removed another key decision-maker at a critical moment.

One would expect also that planners would consider the economic implications of any plan. And that should mean, at the very least, a prominent role for the finance minister in the planning process. In the consideration of war plans, one should expect to find coordination efforts involving political, military, and economic leaders. But it is a rare account that mentions a nation's finance minister. Few accounts reveal any consultation with him as to likely costs, feasibility, and possible difficulties. A finance minister played a decisive role in the Seven Years' War. The conflict ended when Austria's finance minister told Maria Theresa that Vienna had reached its financial limit.[23]

The logical inference, the importance of finance, comes easily but again in some instances that assumption proves inappropriate. At times, some other logic, some other weighting of relevance has prevailed. In 1911, Kaiser Wilhelm's finance and navy ministers reported the impossibility of the projected naval expansion against Great Britain. The monarch's response was that "the question of money plays no role in this." In another conversation on the same subject, Wilhelm declared: "There is enough money available. The Reich's Treasury does not know what to do with all that money . . ."[24]

[22] For the experience of France and Italy, see Hamilton and Herwig, *Origins*, pp. 231, 362; also, Christopher Andrew, "France and the German Menace," in May, ed., *Knowing One's Enemies*, pp. 130–1.

[23] Christopher Duffy, *The Army of Maria Theresa: The Armed Forces of Imperial Austria, 1740–1780* (Vancouver and London, 1977), p. 124.

[24] Volker R. Berghahn and Wilhelm Deist, eds., *Rüstung im Zeichen der wilhelminischen Weltpolitik. Grundlegende Dokumente 1890–1914* (Düsseldorf, 1988), p. 334; and Walter Görlitz, ed., *Der Kaiser . . . Aufzeichnungen des Chefs des Marinekabinetts Admiral Georg Alexander von Müller über die Ära Wilhelms II* (Göttingen, 1965), p. 90.

Effective war planning would, as indicated, require the involvement of economic elites. But here too one finds the unexpected, the infrequent presence of leading industrialists or bankers. Although difficult to establish, one explanation for this planning failure might be bias. Military leaders appear to have a dislike (or disdain) for "economics" and as a result they tend to by-pass the issues involved. One leading nineteenth-century strategist published a 541-page book arguing the influence and importance of sea power upon history. It was a work of advocacy, making the case for a large and effective navy. The work had major impacts, in the United States and, even more, in Germany and Japan, impacts that might well be termed "world historic." The author, Alfred T. Mahan, signaled the importance of colonies as bases for a navy. "Such colonies," he declared, "the United States has not and is not likely to have." The reason given by Mahan was a sharp difference in the concerns of the two leadership groups, the military and the "trading people." On the following page, Mahan broached the question of defending the nation's seaports from "capture or contribution," mentioning the "great cost" involved. His next sentence declares that: "So far as this question is economical, it is outside the scope of this work. . . ."[25]

Mahan's description of the attitudes of "trading peoples" is probably accurate. The disdain for war, as seen, was a long-standing theme in the liberal worldview. The "hawks" of 1898, including Theodore Roosevelt, were well aware of the business opposition to involvement in a war with Spain and condemned them for their lack of patriotism. At that time, Mark Hanna, a prominent businessman and also President William McKinley's campaign manager, declared that "War is just a damn nuisance."[26] And, as seen, businessmen in at least three of the European powers in 1914 opposed the moves toward war. In September 1914 Jack Morgan, head of the famous banking house, saw the war as portending "the most appalling destruction of values in securities which has ever been seen in this country."[27] In 1939, as war again threatened, another businessman, Joseph Kennedy, then ambassador to Great Britain, was

[25] Alfred T. Mahan, *The Influence of Sea Power Upon History, 1660–1783* (Boston, 1925 [1890]), pp. 84, 83.

An account of United States naval expenditures for 1904 reports the division of opinion between two relevant elites – "The fact that all responsible naval opinion in the country was agreed upon the necessity of these distant [western Pacific] bases made no impression upon Congress." From Gordon Carpenter O'Gara, *Theodore Roosevelt and the Rise of the Modern Navy* (Princeton, 1943), pp. 28–29.

[26] Thomas Beer, *Hanna* (New York, 1929), p. 206.

[27] Ron Chernow, *The House of Morgan* (New York, 1990), p. 185.

"of one mind" with his close friend, Neville Chamberlain, on the importance of avoiding war. A friend described Kennedy's views as follows: "War was irrational and debasing. War destroyed capital. What could be worse than that?"[28] The basic dispositions of business and military elites, in short, led them in opposite directions. For those two leadership groups discussion of a possible war would, under ordinary circumstances, be rather difficult.

A more important reason for the exclusion of business elites from discussion of war plans would again be the need for secrecy. Any such discussion would require the revelation of the current plans and imminent intentions. Even if only a small number of leading industrialists and financiers were involved, it would be likely that some details of the plan would be passed on to others, to family members, colleagues, or friends and even that some of this information would come to the attention of enemies. Some participants might tell others directly – "in strictest confidence" – of the ongoing planning. Some might communicate the new knowledge indirectly. Those possessing such "inside information" would have a strong incentive to withdraw investments from the prospective enemy nations, a move that would be detected by any alert intelligence agency.

Economic considerations impact on planning at every point, big and small. Put differently, every aspect of the process has a cost. In July 1914 the American vice-consul general in Budapest, Frank Mallett, "noted the Austrian mobilization on the Serbian frontier and the commandeering of horses by the Austrian army." On 13 July he wrote a dispatch, the first report telling of the imminent war between Austria and Serbia. But the dispatch was sent by ordinary post and did not reach the State Department until 27 July. Here is the explanation:

> To have sent it by cable would have drawn a rebuke from the department, the warning so frequently given to ministers and consuls whose telegrams were long. Mallett's letter contained about 150 words. The men in the cable and telegraph room downstairs at the department objected to long messages because of the time it took to decipher them; besides, each word cut into the consular service's budget.[29]

The neglect of economic realities, the financial costs of war, is remarkable. Consider the following: a general mobilization would remove all

[28] Richard J. Whalen, *The Founding Father: The Story of Joseph P. Kennedy* (New York, 1964), p. 234.

[29] Cited in Rachel West, *The Department of State on the Eve of the First World War* (Athens, Ga., 1978), p. 133.

able-bodied males, those of ages 18 to 40, from the labor force. Virtually every productive process, be it the manufacture of food, clothing, furniture, hardware, or steel, would be interrupted. Even more serious in August 1914, farm laborers would be shifted to the military, just as crops were about to be harvested. It follows that virtually all consumer goods would suddenly be in short supply bringing another obvious consequence, rising prices. This would occur at a time when family incomes were suddenly reduced. French officials did study this question carefully. The effect on food supply was their greatest concern.

In addition, arrangements would have to be made for the shifting of production from peacetime goods to those needed in war. Some of those changes would be relatively easy. The shift from civilian to military clothing, for example, would present relatively small difficulties. Given a largely female labor force in the clothing industry, the problem could be handled. But the shift in other product lines, all branches of steel manufacture for example, would pose much greater difficulties. Major plant conversions would be required, for example, to shift from the production of steel rails and locomotives to artillery, tanks, and barbed wire. For Germany, one major industry, ocean shipping, would be shut down entirely. Ships in home ports could not leave; all ships at sea were effectively lost.[30] Some writers have argued that German leaders went to war in the pursuit of imperialist goals – it was a "grab for world power."[31] But without ships at sea, naval or transport, Germany effectively lost all of its overseas possessions within months of the outbreak.

Another factor that might block otherwise expected critical interventions is a normal social psychological tendency, one found in all settings, that being the problem of conformity, also called groupthink. One very talented decision-making group has been described as "the best and the brightest," this consisting of an American president, John F. Kennedy, his cabinet, and some top advisors. At one important meeting, Central Intelligence Agency representatives informed the group of a plan for landing a force in Cuba at the Bay of Pigs. After the disaster, Arthur

[30] For the reaction to the news of war by Germany's leading shipping magnate, Albert Ballin of the Hamburg-America Line, see Hamilton and Herwig, *Origins*, pp. 486–7; and Lamar Cecil, *Albert Ballin: Business and Politics in Imperial Germany, 1888–1918* (Princeton, 1967), pp. 206–8, 212.

[31] Fritz Fischer, *Griff nach der Weltmacht. Die Kriegszielpolitik des kaiserlichen Deutschland 1914/18* (Düsseldorf, 1961); Imanuel Geiss, *German Foreign Policy 1871–1914* (London, 1976); Volker R. Berghahn, *Europe in the Era of Two World Wars* (Princeton, 2006).

Schlesinger described the decision making as follows: "...the massed and caparisoned authority of [Kennedy's] senior officials in the realm of foreign policy and defense was unanimous for going ahead.... Had one senior advisor opposed the adventure, I believe that Kennedy would have cancelled it." But, Schlesinger added: "No one spoke against it."[32]

Another social-psychological factor is mindset. An elite whose task is the defense of the nation will focus on threat perception. They will collect "intelligence" about other states, their military potential, and the intentions of their leaders. Much of that information would be rather sketchy, some of it of questionable value. The conclusions drawn, with rare exception, depend heavily on inference or worse, on flimsy guesswork. Not too surprisingly, the readings and responses sometimes suggest paranoid tendencies. Germany's political and military leaders, for example, were concerned with the problem of "encirclement." The threat was summarized by Schlieffen in his last statement which depicted Germany and Austria standing unprotected in the middle of Europe, surrounded by the ramparts and trenches of their dedicated enemies, France, Britain, and Russia. "At the given moment," he continued, "the gates will be opened, the drawbridges let down, and armies of millions of men will pour into Central Europe across the Vosges, the Meuse, the Königsau, the Niemen, the Bug, and even across the Isonzo and Tryolean Alps in a wave of devastation and destruction." Simultaneously, another threat – "the red specter" – was thought to be hovering in the background. But Dieter Groh's massive study, based on German and Berlin police records, shows that Chancellor Theobald von Bethmann Hollwegg and other government leaders knew precisely how "patriotic" the Social Democratic party was in 1914 and that it posed no vital threat. That is why they, like the French, cancelled all plans to arrest leading socialists in 1914.[33]

Business elites generally operate with a different, more open, mindset, their prime concern being the discovery and development of commercial

[32] Cited in Irving L. Janis, *Victims of Groupthink: A Psychological Study of Foreign-Policy Decisions and Fiascoes* (Boston, 1968), Ch. 2, "A Perfect Failure..." (the quotation, p. 39). For much more, see Arthur M. Schlesinger, Jr., *A Thousand Days: John F. Kennedy in the White House* (Boston, 1965), Ch. 10.

See also: Geoffrey Regan, *Someone Had Blundered. A Historical Survey of Military Incompetence* (London, 1987); and Eliot A. Cohen and John Gooch, *Military Misfortunes: The Anatomy of Failure in War* (New York, 1990).

[33] Alfred von Schlieffen, "Der Krieg in der Gegenwart," BA-MA, Nachlass Schlieffen, N 43, Vol. 101; and Dieter Groh, *Negative Integration und revolutionärer Attentismus: die deutsche Sozialdemokratie am Vorabend des Ersten Weltkrieges* (Frankfurt/Main, 1973).

opportunities at home and abroad. Hugo Stinnes, one of Germany's top business leaders, had holdings in Luxembourg, French Lorraine, Normandy, Belgium, Wales, St. Petersburg, and the Ottoman Empire. One aspect of his vast business empire involved the transport of English coal on German ships for delivery to St. Petersburg for use by the Russian navy. A war involving a sweep through Luxembourg, Belgium, and northern France would obviously damage his interests. Not surprisingly, he showed little interest in militarism and imperialism. His business policies, accordingly, were "geared toward a long period of peace." In the summer of 1913 his hope was that the people in the Balkans would "settle down." At that point he acquired new mining interests in Yorkshire and Nottinghamshire.[34]

Several other fundamental considerations should be added to this review of "the basics." First, an elementary one: would the war be short or long? There was a general expectation, or hope, that it would be short.[35] But there was also a suspicion, a fear that it might be long. Prior experience provided little guidance. After the fact, one might easily point to the Crimean War (1853–56) as an appropriate "model." Some commentators did point to the American Civil War (1861–65) as the most relevant experience. But one could also point to contrary experience, to three brief wars, Otto von Bismarck's wars of German unification, against Denmark in 1864, against Austria in 1866, and against France in 1870–71. Western Europe had been spared such ventures for more than four decades, from 1871 to 1914, and hence had little direct immediate experience. One might look at the Russo-Japanese War (1904–1905), but that was easily seen as atypical, a curious aberration ending with the defeat of a frail and tottering empire.

Some conflicting lessons were seen elsewhere. Would the impending European conflict be quick and easy as with the Spanish-American War (1898)? Or would it be long and hard-fought as in South Africa's Boer War (1899–1902)? But against the Spanish-American case there was the experience of another struggle, the subsequent long and bitter campaign in the Philippines. Given the uncertainties, there was clearly a need for at least two plans, for the long- and the short-term possibilities.

[34] Hamilton and Herwig, *Origins*, pp. 490–2.
Our intent with these comments on mindsets is to signal tendencies within the two elite groups. These are not categorical statements. The claims are testable with some fragments of evidence provided here and more to come in the following chapters.

[35] See Lancelot L. Farrar, Jr., *The Short-War Illusion: German Policy, Strategy & Domestic Affairs, August-December 1914* (Santa Barbara, 1973).

What of its size? Would the war be large, middling, or small? Most decision-makers in 1914 assumed it would be large, one with much involvement. The Austro-Hungarian, Russian, and German efforts would begin with full mobilization of their forces, an option taken for granted from an early point. That would still leave the short versus long question. If not successful in the first "round," what were the plans for the longer struggle?

These examples provide but a brief introduction to the subject, to the unexpected behaviors found in connection with the subject of war planning.

Like our previous volume, *The Origins of World War I*, in this work we are arguing against a style of historical writing, one that focuses on "big" causes. Some writers report the presence of various "forces" that somehow brought on the war. Among the "gigantic forces" said to have been "in motion" are those of nationalism, social Darwinism, militarism, imperialism, and "the alliances."[36] Many such discussions make use of reification, treating those abstractions as agencies that somehow caused the result. The authors of one leading textbook, for example, open their discussion of World War I with several such declarations, including: "Somewhere before 1914 Europe went off its course . . . Europe stumbled in 1914 into disaster . . . Europe went astray."[37] The appropriate response to all such reifications would be emphatic denial: *Europe* did no such thing. Europe was not "an actor."

Another statement from the same textbook declares that "After 1870 Europe lived in a repressed fear of itself."[38] Here the authors suggest they know the hidden feelings present among the Continent's masses. The discovery of mass sentiments is a difficult task at any time, even when aided by the best tools of opinion research. The claim about repressed fears in the nations of Europe prior to 1914 should be recognized as, at best, an untested hypothesis. It is a declaration that, long after the presumed fact, could not possibly be verified.

A sub-heading introduces another of the "big causes" – "Rival Alliances – Triple Alliance versus Triple Entente." Three great powers of Europe, Austria-Hungary, Germany, and Italy, are said to be allied against another opposing three, Russia, France, and Great Britain. The

[36] See Hamilton and Herwig, *Origins*, pp. 16–41.

[37] R. R. Palmer and Joel Colton, *A History of the Modern World since 1815* (8th edition, New York, 1995), p. 695.

[38] Loc. cit.

terms of the Triple Alliance are described briefly – and inaccurately – as follows: "if any member became involved in war with two or more powers its allies should come to its aid by force of arms." That alliance, however, like most such agreements, contained a very important specification: aid was required only if that member's involvement was unprovoked. Austria-Hungary's attack on Serbia was seen, by most of Europe's leaders, as a very provocative move, one that after Austria-Hungary's bombing of Belgrade brought Russia's general mobilization. That provocation provided the justification for Italy's leaders, for their decision to remain neutral.

The other term, Triple Entente, is a misnomer. It was a Double Entente, something indicated six pages later where it is stated, accurately, that Great Britain "was bound by no formal military alliance."[39]

In all six major powers, the decisions for (or against) war were made by coteries, their choices dependent on readings of events in July and August of 1914. A few days after the assassination of Franz Ferdinand, Austria-Hungary's leaders decided to attack Serbia and, recognizing the likelihood of Russian intervention, sent a mission to Berlin to gain Kaiser Wilhelm's assurance that Germany would back them. Austria-Hungary's arrangements with Germany were made prior to any attack by another power. The imminent threat, moreover, was Russia, not "two or more powers..."

That reading, alliance determination, fails in another respect. It suggests that the leaders of Europe's great powers were honorable men, all "bound by" those alliance provisions, obligated to follow them regardless of circumstances and likely consequences. That assumption is challenged, first, by the Austrian move to operate outside of the alliance rules. Second, the German leaders' choice of war was not stipulated by any alliance obligations; that choice also was outside of, entirely independent of the treaty requirements. And third, Germany's first action in the war involved a flagrant violation of another treaty. In 1839, the major European powers, including the leaders of Prussia, Germany's predecessor, had signed a treaty declaring their obligation to protect Belgium's neutrality. The Germans were not alone in this respect. Earlier, French war planners had also contemplated a move into Belgium. But they did not follow through. French political leaders told them that such a move would lose France the support of Great Britain and Italy.

[39] Ibid., pp. 696, 702.

Declarations about the impacts of "powerful" forces, in short, are mistaken. All such statements need translation, a specification of meanings. Decisions for war were made by coteries of leaders. In most cases, fewer than a dozen persons were directly involved in those decisions.

These observations apply also to statements about war plans and mobilization schedules. No plan "dictated" reactions; no mobilization schedule "called for" some action. Plans and schedules are inert objects, things that by themselves "do" nothing. Planning and implementation efforts, like the choices of war, were also the work of coteries, of diverse decision-making groups. And those planning processes, as will be seen in the following chapters, were also subject to a wide range of individual whim and contingency.

2

Austria-Hungary

Günther Kronenbitter

In 1914, the Habsburg Monarchy had to face the most difficult strategic situation of any of the Great Powers. Home to people who spoke eleven different officially recognized languages and without an ethnic group dominating state and culture as Russians and Turks did in the tsar's and the sultan's empires, Austria-Hungary was unique. The Monarchy suffered from ethnic conflicts and was almost surrounded by potential foes. In terms of military power, Russia, competing with Austria-Hungary for influence on the Balkan Peninsula, was the most dangerous of them. Since the regicide of King Alexander Obrenovich in 1903, Serbia also challenged the Habsburg Monarchy's supremacy in the area. The Serbian army did not look very menacing to Austria-Hungary's forces until 1912–13 when the Serbs proved to be superior in fighting power to Ottoman and Bulgarian troops. Moreover, mounting unrest among the Serbian population in Bosnia and the southern parts of Hungary fostered fears of an uprising in case of a war between the Habsburg Monarchy and Serbia or Montenegro, another South-Slav neighbor. Tiny Montenegro under King Nikola Petrovich Njegosh tried to expand its influence and forged dynastic ties with Russia and Italy.

Italian nation-building in the nineteenth century had come at the expense of Austria, but there were still some territories with an Italian majority under Habsburg rule. Italy, like Germany, was Austria-Hungary's partner in the Triple Alliance, but relations were usually strained. In the last years of peace, the reliability of Romania, which had acceded to the Triple Alliance, could no longer be counted upon.

Nationalism fueled tensions between Austria-Hungary and Serbia, Montenegro, Italy, Romania, and – with regard to eastern Galicia – even

Russia. The multiethnic Habsburg Empire seemed extremely vulnerable to any alliance between nationalist groups and foreign governments. Social-Darwinist patterns of perception lent plausibility to claims that the laws of history and nature would sooner or later turn potential opponents into a deadly threat to the Monarchy.

At the same time, the loyalty of the Czechs, Italians, South Slavs, Romanians, and Ukrainians serving in the Habsburg forces seemed questionable in case of a war. The conflict with Serbia and tensions with Russia, just like the uneasy partnership with Italy and Romania, bears witness to the close relation between Austria-Hungary's domestic and international problems. Any major armed conflict could spur the disintegration of the Monarchy. Even if there were no signs of an imminent attack on Austria-Hungary, the widespread fear of a shifting balance of military power and the unstable domestic situation made governing circles in Vienna and Budapest anxious to avoid a multi-front war. That was the political context of strategic and operational considerations in the General Staff and it shaped the mission of Austria-Hungary's military intelligence.

Threat Perception and the Limits of Intelligence

The organizational framework of Austria-Hungary's military intelligence differed from that of Germany. In the Intelligence Office (*Evidenzbüro*) in Vienna, separate "groups" or departments for each of the main military powers in Europe and for all of Austria-Hungary's neighbors gathered and processed information from a wide range of sources, most of them publications and the reports of military attachés. Espionage and counterintelligence were the preserve of a specialized group, the *Kundschaftsgruppe*. In comparison with the German Sektion IIIb, military espionage in the Austro-Hungarian army enjoyed less independence. The head of the *Kundschaftsgruppe* reported to the chief of the *Evidenzbüro*, who in turn controlled the flow of information to the other offices and the chief of the General Staff.

In the last years before the war, there was talk of a root-and-branch reform of this system. Closer to the German model, the new structure would put the separate groups of the *Operationsbüro* in charge of collecting and processing the information they needed. The head of the *Kundschaftsgruppe*, the Habsburg monarchy's spymaster, would report directly to the chief of the General Staff or his deputy. The *Evidenzbüro* would have been dissolved and most of its personnel transferred to the various groups of the *Operationsbüro*. Had this reform been implemented, one hierarchical layer would have been eliminated.[1]

Emulation of the German model was a widespread concern of many reform-minded officers in the Austro-Hungarian army, not just in the General Staff. Nevertheless, the example of the German Sektion IIIb would not have been enough to trigger a debate about the reform of the intelligence branch. What made the *Evidenzbüro* so vulnerable to criticism was the most spectacular scandal in the history of the army: the Redl-affair. In 1913, the former head of the *Kundschaftsgruppe*, Colonel Alfred Redl, was arrested on charges of treason and eventually committed suicide. He had provided several foreign intelligence services, notably the Russian military espionage office in Warsaw, with secret information. Because of his foible for luxury and the money he needed to support his lover, an army officer, Redl had become a traitor. This was bad enough for the *Evidenzbüro*, but what made the scandal even more depressing was the fact that without the helping hand of the German Sektion IIIb, Redl would not have been caught.

The Redl-affair dealt a severe blow to the prestige of the *Evidenzbüro* and the General Staff. Redl's suicide, instigated by the department heads of the General Staff, was meant to defend the officer corps' honor, but it also made it easier to cover up the dimensions of his betrayal. Neither Kaiser Franz Joseph nor Archduke Franz Ferdinand, the heir apparent to the throne – not to mention the German General Staff – received a complete and open brief about the probable damage Redl had done to the army's cause.[2] Rumor had it that the Russians in August 1914 made good use of the deployment plans Redl sold them. But there is precious little evidence to support this assumption, since the Austrian General Staff

[1] For the *Evidenzbüro*, see Norman Stone, "Austria-Hungary," in Ernest R. May, ed., *Knowing One's Enemies: Intelligence Assessment Before the Two World Wars* (Princeton, 1984), pp. 37–61; Albert Pethö, *Agenten für den Doppeladler. Österreich-Ungarns geheimer Dienst im Weltkrieg* (Graz and Stuttgart, 1998), pp. 14–24; Günther Kronenbitter, *"Krieg im Frieden." Die Führung der k. u. k. Armee und die Großmachtpolitik Österreich-Ungarns 1906–1914* (Munich, 2003), pp. 233–47.

[2] For Redl and the Redl-affair, see Georg Markus, *Der Fall Redl* (Vienna and Munich, 1984); Pethö, *Agenten*, pp. 227–38; John R. Schindler, "Redl-Spy of the Century?" *International Journal of Intelligence and Counterintelligence* XVIII (Fall 2005). On the Russian perspective and the situation in 1914, see Bruce Menning's chapter on Russia. Secret documents concerning the Redl case were discovered by the author of this chapter in a sealed envelope, hidden among files on espionage-minutiae, in the Kriegsarchiv (War Archive) in Vienna. They give reference for the Viennese General Staff's policy to deliberately mislead the Germans about the damage done by Redl. See Kronenbitter, "*Krieg im Frieden*," pp. 236–7.

had altered deployments in Galicia by summer 1914. The most obvious and tangible result of Redl's betrayal was the breakdown of Austria-Hungary's spy network in the tsar's empire. Thanks to Redl, Russian counterintelligence could disrupt Vienna's espionage efforts. From 1903 to 1906, in a time of relative détente between Russia and the Habsburg Monarchy, cost-cutting measures had almost brought Austrian espionage in the East to a halt. The Redl-affair undermined the revival of Austrian intelligence operations inside Russia.

The rather underwhelming performance of Austria-Hungary's spy circles forced the *Kundschaftsgruppe* to rely on cooperation with the German Sektion IIIb on the one hand, and on anti-Russian Polish underground movements on the other. The Poles, who were striving for independence from St. Petersburg and who hoped for a clash between Russia and its neighbors to the West, had an interest in painting Russian military preparations in stark colors. In the Balkan crises of 1912–13, when war between Russia and Austria-Hungary seemed imminent, the perception in Vienna of Russia's military build-up as extremely aggressive and menacing was partly due to these sources. Sektion IIIb had to soothe the nerves of Berlin's political and military leadership as the alarming news from Vienna reached Berlin. The Germans realized that the view from Vienna was far too pessimistic. But in general, Sektion IIIb and the *Kundschaftsgruppe* cooperated smoothly and quite efficiently. This is one of the reasons why the Austro-Hungarian General Staff had a reasonably good understanding of Russia's war preparations.

More important probably was the fact that the *Evidenzbüro* could gather a lot of valuable information from open sources, such as publications, parliamentary debates, budget details, or the reports of military attachés and the embassy in St. Petersburg. The build-up of new transport capacities in Russia, one of the decisive aspects of war preparations in those days, could be studied by officers traveling the country in civilian clothes, or by locals, many of them Jews, passing information on to the intelligence officers of the corps commands in Galicia. Compared to the budget of the Foreign Ministry's intelligence service, the *Evidenzbüro* was grossly under-funded.[3] All in all, the hard facts of Russian military

[3] Conrad and Schemua used the crises of 1912–13 to achieve a substantial rise of the budget of the *Kundschaftsgruppe*. But financial problems never lost their negative effects on intelligence efforts. It took the General Staff several months to fork out the relatively minuscule amount of money needed to reconnoitre the railway links in northwestern Serbia in 1914. Österreichisches Staatsarchiv-Kriegsarchiv, Vienna (hereafter, ÖstA, KA), Gstb 1914 69–1.

preparations – troop numbers, the equipment and maintenance of fortresses, railway capacities, and the like – were a much better basis for Austria-Hungary's deployment plans than information on strategic or operational planning.

Basically, the same held true for the Habsburg army's other potential foes. Intentions were hard to read but capabilities could be scrutinized to great effect. Austrian intelligence monitored the changes in Italy's transport system and the building of new fortifications close to the border. Because the Italian fleet was considered to be the yardstick for Austria-Hungary's naval build-up, the intelligence office of the Habsburg navy focused on Italy's sea power. On the Balkan Peninsula, Serbia became the most important target for Vienna's espionage, but the Ottoman Empire and Romania were also of great interest. To an even higher degree than in St. Petersburg and in Rome, the military attachés in the Balkan capitals and in Istanbul, assisted by the consuls in the area, provided the General Staff with background information.

Nevertheless, Austrian intelligence and diplomacy failed to realize the dramatic shift in alliance patterns and the formation of a Balkan League in 1912. The outbreak of the First Balkan War took civilian and military leaders in Vienna by surprise. The shock of 1912 paved the way for new intelligence efforts. The number of military attachés in southeastern Europe was increased, and more money for espionage became available. At the same time, the Balkan wars of 1912–13 made it difficult to follow the rapidly changing military situation on the peninsula.

But this was not the biggest challenge to the General Staff's intelligence office. Rather, the assessment of the fighting power of Austria-Hungary's neighbors proved to be much harder, and not just in the case of Serbia or Montenegro. The Bulgarians in 1912 and the Serbs in both Balkan wars did surprisingly well. The traditional condescending views of Balkan armies gave way in Vienna to a more nuanced assessment of their fighting effectiveness.[4] All in all, the picture painted by the *Evidenzbüro* was quite accurate – and bore witness to Austria-Hungary's deteriorating strategic situation.

The General Staff's Monopoly on War Planning

Among the Great Powers of Europe in 1914, the Habsburg Monarchy looked like the odd man out: as a multiethnic empire, cast in a

[4] Stone, "Austria-Hungary," pp. 43–4; Kronenbitter, "*Krieg im Frieden*," pp. 434–5.

constitutional arrangement of bewildering complexity, it relied on tradition and convenience rather than on nationalism or visions of progress and overseas expansion. Since 1867, the Compromise (*Ausgleich*) was the constitutional basis for the Habsburg Monarchy's political structure. Austria-Hungary was composed of two separate states, Hungary and the hereditary Austrian Habsburg holdings. Each had its own parliament and government, responsible for most aspects of domestic policies and a great deal of economic affairs. The monarch, who was both king of Hungary and emperor of Austria, appointed the prime ministers and could veto legislation, providing the strongest tie between both states. The common army and navy were run by the Ministry of War and paid for by the Common Ministry of Finance. Austria-Hungary's diplomacy was steered by the foreign minister, who presided over the Common Ministerial Council, the highest-ranking body of decision-makers in Austria-Hungary. The prime ministers of Austria and Hungary, often joined by their ministers of finance and the three common ministers, discussed foreign affairs, budgetary questions concerning the armed forces, and tariffs, to name but the most important. In the end, the consent of the Council, of both parliaments, and of their non-permanent committees (the so-called "delegations") was essential for most resolutions. On war and peace, the parliaments had no say. Foreign affairs and the command of the armed forces were the sole prerogative of the monarch, who was advised by the common ministers and the prime ministers. Any decision concerning the army and navy that was deemed important had to get the sanction of the emperor/king. Thus it was quite appropriate to call the common army the Imperial and Royal (k. u. k.) Army.[5]

In 1848–49, the army had saved the Habsburgs from revolutionary and nationalist movements, spilling the blood of invading troops from the Piedmont, insurgents in Bohemia and Vienna, and Hungarians fighting for independence. Franz Joseph, who had become emperor at the height of revolutionary turmoil in December 1848 when he was just 18 years old, never forgot the lessons of 1848–49: when push came to shove, the armed forces had to be relied on as the backbone of Habsburg rule. In the wake of revolution, the army was considered first and foremost as a tool of control and repression in the Habsburgs' lands. This focus on policing and

[5] For general histories, see Robert A. Kann, *A History of the Habsburg Empire 1526–1918* (Berkeley, 1974); see *The Last Years of Austria-Hungary: Essays in Political and Military History, 1908–1918*, ed. Mark Cornwall (Exeter, 1990), for the last six years of the prewar period; Samuel R. Williamson, Jr., *Austria-Hungary and the Origins of World War I* (New York, 1991).

financial constraints was partly responsible for shortcomings in Austrian war preparations. Dismal diplomacy and strategy, flawed operational leadership, and a widespread reluctance to adapt new weaponry and tactics led the Austrian armies to defeat in 1859 and 1866. After 1859, Franz Joseph, who had commanded his troops at the Battle of Solferino, steered clear of operational or tactical decisions.

Archduke Franz Ferdinand played a prominent role in the military and was appointed Inspector General of the armed forces in 1913. His military chancellery offered him an institutional platform from which to launch political campaigns. The archduke and his staff watched over domestic politics and tried to make or break careers in government, the civil administration, and the military according to Franz Ferdinand's judgment on the loyalty and reliability of office-holders and seekers. He desperately tried to reduce the influence of the Hungarian political elite, of liberals, of Protestants, and of Jews. Resentment, not reform, shaped Franz Ferdinand's political mindset, but his reactionary leanings made him anxious to avoid an armed clash with Russia. In domestic and military affairs, the archduke was less farsighted and it proved difficult for the top brass to keep his support.[6] After his assassination at Sarajevo in June 1914, there was no one left in the Monarchy's leadership circle to veto a confrontation with Russia.

The minister of war, Franz von Schönaich, was one of several senior leaders who fell out of the archduke's favor shortly after being appointed to a leading position. In Franz Ferdinand's eyes, Schönaich was willing to strike a deal with Hungarian leaders in order to get political backing for more recruits and a bigger budget. Finally, the so-called "Belvedere circle" around the archduke brought Schönaich down in 1911. Neither Schönaich nor his successors – Moritz Ritter von Auffenberg and Alexander Ritter von Krobatin – tried to exert leverage on war planning undertaken by the General Staff. Franz Joseph had to sanction all of Austria-Hungary's war plans, but he did not interfere in the process. For his part, Franz Ferdinand lacked the skills, the discipline and the time to grapple with the details of war planning. But his chancellery analyzed the plans and at least on one occasion even proposed modifications. Neither the

[6] For Franz Ferdinand and his military chancellery, see Robert A. Kann, *Erzherzog Franz Ferdinand Studien* (Vienna, 1976); Samuel R. Williamson, Jr., "Influence, Power, and the Policy Process: The Case of Franz Ferdinand, 1906–1914," *Historical Journal* 17 (1974), pp. 417–34.

commander of the Austrian-Hungarian navy nor the General Staff took an interest in coordinated planning for land and naval warfare beyond minor landing operations or coastal defense in the Adriatic Sea. In short, within the military apparatus, there was no effective control of the General Staff's work. The monarch and the heir-apparent simply did not make it their business to question war plans. Furthermore, the Foreign Office and the Common Ministerial Council were officially briefed by the General Staff only in times of crisis. Thus, the General Staff monopolized war planning in the Habsburg Monarchy and its chief played a pivotal role in Austrian war preparations.

Friedrich von Beck-Rzikowsky had emulated the Prussian model when he established the General Staff in Vienna as the center of war preparations. In 1906, Beck was replaced by Franz Conrad von Hötzendorf, the commander of an infantry division in Tyrol, a province bordering Italy, one with a sizeable Italian-speaking minority. Conrad had built his reputation as the leading expert on tactical matters in the k. u. k. army on the basis of a number of widely-received publications. A prolific writer, he produced thousands of memoranda, reports, and letters in order to propagate his views on the need to reform the army as soon and as thoroughly as possible, and to call for preventive wars against Italy and Serbia as the only viable solutions to Austria-Hungary's deteriorating strategic situation.[7] Conrad had a firsthand knowledge of the southwestern borderlands and was convinced that sooner or later Italy would try to acquire further territory at the expense of the Habsburg Monarchy. He considered it his duty as the emperor's loyal servant to speak out against diplomatic efforts to avoid confrontations with Rome or Belgrade.

During the Bosnian Crisis of 1908–09, the chief of the General Staff and Austria-Hungary's foreign minister, Alois Lexa von Aehrenthal, got along rather well for quite a while. When Conrad realized that Aehrenthal would not launch a war against Serbia after the government in Belgrade had given up its resistance against the annexation of Bosnia-Herzegovina, he risked open conflict with the Foreign Ministry. Conrad's attempts to talk the monarch into an offensive war against Italy in 1911 backfired and Conrad lost his job. Stubborn yet eloquent, Conrad fought for his convictions relentlessly. It would be wrong to deny the important role his

[7] Franz Conrad von Hötzendorf, *Aus meiner Dienstzeit, 1906–1918*, 5 vols. (Vienna, Leipzig and Munich, 1921–25); *Private Aufzeichnungen*, ed. Kurt Peball (Vienna and Munich, 1977).

character traits played in the security policy of Austria-Hungary in the prewar decade.

Conrad's successor as chief of the General Staff, Blasius Schemua, had less self-confidence and refrained from an open confrontation with the new foreign minister, Count Leopold Berchtold. Schemua called for a more assertive foreign policy, which he couched in social Darwinist language. Unlike Conrad, Schemua had a soft spot for theosophic and *völkisch* ideas. Like Conrad, Schemua was deeply pessimistic about the Habsburg Monarchy's survival in the long run. Imperialist expansion on the Balkan Peninsula and war against Italy, both soldiers argued, would slow down Austria-Hungary's perceived decline and might give it a new lease on life – at least for a while.

Conrad returned to his former position after one year in "exile." In the acute international crisis triggered by the First Balkan War, it made sense to ask Conrad to take over again not only because of his unquestionable energy, but also because the chief of the General Staff might at any moment be called upon to implement the deployment and operational plans produced in years before. In addition, Conrad's comeback helped to stimulate the General Staff. What really made Conrad different was his ability to enthuse younger officers. Even when they became critical of their master's plans and his later performance as commander in the Great War, most of them maintained respect for his determination and will power. There were cliques within the military leadership, but most of them had no other agenda than to foster their careers. Very few of the high-ranking officers ever challenged Conrad's decisions on strategic or operational questions.

The heart of the General Staff, holding the decisive position in the planning process, was the Operations Office, or *Operationsbüro*. When Conrad succeeded Beck, Heinrich Ritter von Krauss-Elislago headed the *Operationsbüro*. Krauss ensured some continuity in the *Operationsbüro*'s planning, but Conrad never really appreciated him. In 1910, Krauss was replaced by Joseph Metzger, an extremely loyal follower of Conrad, much like Franz Höfer, the chief's deputy. Metzger became the mentor of a group of younger General Staff officers. They, like all of their colleagues, were graduates of the War Academy, the *Kriegsschule*, and survived a rigorous process of selection before joining the General Staff. Most of them were from lower middle-class backgrounds; there were many sons of k. u. k. officers among them. Conrad's energy, his laid-back views on social etiquette, and his benevolence made him popular with the up-and-coming young officers. The chief of the General Staff even discussed sensitive issues such as the political and strategic situation with them.[8]

Political Flexibility and Operational Rigidity: War Cases and Deployment Plans

The General Staff had to come to terms with the Monarchy's unfavorable geo-strategic position. When Conrad succeeded Beck, he inherited plans for wars against Russia, Italy, and Serbia in the files of the Operations Office. The new chief of the General Staff gave an order to revise war plan "I" (Italy) because he knew that war theater quite well and had drafted plans for deployment and operations there during his tenure as commander in Tyrol. In addition, he considered a preventive war against Italy necessary and imminent. Quite soon, plans for war cases "R" (Russia) and "B" (Balkan, that is, Serbia or Serbia and Montenegro) were also recast. In 1913–14, deployment plans in case of a war against Romania were worked out.

The most striking characteristic of Austria-Hungary's war plans on the eve of Sarajevo was not the sheer number of potential foes but the urgent need to keep war preparations flexible. Under Beck, the political situation allowed for one-front plans, but even then a two-pronged attack on Italy and on Serbia was prepared. After 1906, a war on one front seemed to be an unlikely scenario. War against Italy might provoke Serbia to intervene and vice versa. Russia was still recovering from its defeat in 1904, but it might join the fray in case of an Austrian war against Italy and/or Serbia. War on two or even three fronts had to be taken into account by Conrad and his staff. Given the limited military resources of the Habsburg Monarchy, focusing first on one opponent and then, after decisive victories, swirling the bulk of the army around against the other, was seen as the only way out of the strategic impasse.

To make life even harder for the planners at the General Staff, the *Operationsbüro* had to prepare for an intervention by additional powers

[8] For Austria-Hungary's top brass on the eve of World War I, see Kronenbitter, "*Krieg im Frieden*," pp. 17–77, which is based on published memoirs, papers of general staff officers in the *Kriegsarchiv* and the transcripts of the diaries of Rudolf Kundmann and – for the first time – of Franz Putz, a Conrad aide-de-camp. For Conrad, see Lawrence Sondhaus, *Franz Conrad von Hötzendorf: Architect of the Apocalypse* (Boston, Leiden and Cologne, 2000); August Urbanski von Ostrymiecz, *Conrad von Hötzendorf. Soldat und Mensch* (Graz, Leipzig and Vienna, 1938). For Beck, see Scott Lackey, *The Rebirth of the Habsburg Army: Friedrich Beck and the Rise of the General Staff* (Westport, Conn., and London, 1995).

that would not take place right away, but at a time when the deployment of k. u. k. forces had already begun. It was unlikely that Austria-Hungary would have to fight on one front only, but it was also improbable that it would not be supported by at least one of its allies. General Staff officers in Vienna thus had to draft war plans with the highest degree of flexibility – a tall order, much more demanding than the tasks of their colleagues in Paris, Berlin, or St. Petersburg.

Conrad tried to combine political flexibility with thorough planning, but there was no way to square the circle. Since the Bosnian crisis of 1908–09, a two-front war against Serbia and Russia emerged as the most probable scenario, but the uncertain political constellation made it all but impossible to predict the timing of Russia's intervention. For purposes of mobilization and deployment, Austria-Hungary's army was split into three striking forces. Of a total of 48½ infantry divisions, about eight would form Minimal Group Balkan (Fifth and Sixth armies) and defend the South Slav provinces or join an attack on Serbia and Montenegro. The bulk of these troops belonged to the two corps in Bosnia-Herzegovina and Dalmatia (XV, XVI) and to the Croatian corps (XIII). In case of an offensive against Serbia, Minimal Group Balkan would be reinforced by the forces of B-Group (Second Army), drawn from two Hungarian (IV, VII) and two Bohemian (VIII, IX) corps. Ahead of these troops, A-Group with at least 28 infantry divisions would be concentrated at the Russian or the Italian border, depending on the political situation. B-Group could be directed to the Balkan theater first and moved to the northeastern or southwestern front in case of a Russian or Italian intervention. Or it could be deployed in Galicia or against Italy right from the beginning. This arrangement was completed by a system, introduced in 1909, of mobilizing the corps independently. All of this allowed for a more flexible response to a shifting strategic situation. If B-Group did not get involved in fighting, its units could be redeployed.[9]

But flexibility came at a price: the railway transport of B-Group had to be detained because A-Group deployment was given a higher priority.

[9] For Austria-Hungary's war plans on the eve of World War I, see Graydon A. Tunstall, Jr., *Planning for War against Russia and Serbia: Austro-Hungarian and German Military Strategies, 1871–1914* (New York, 1993), pp. 55–135; Dieter Degreif, *Operative Planungen des k. u. k. Generalstabes für einen Krieg in der Zeit vor 1914 (1880–1914)* (Wiesbaden, 1985), pp. 113–298; Hans Jürgen Pantenius, *Der Angriffsgedanke gegen Italien bei Conrad von Hötzendorf. Ein Beitrag zur Koalitionskriegsführung im Ersten Weltkrieg*, vol. 1 (Cologne and Vienna, 1984), pp. 425–62; John H. Maurer, *The Outbreak of the First World War: Strategic Planning, Crisis Decision Making and Deterrence Failure* (Westport, Conn., and London, 1995), Chs. 2 and 3.

In addition, Conrad conceded that it would be impossible for the Railway Office of the General Staff to prepare detailed transport plans for each of the possible scenarios. The officers in charge of railway transports did their very best to make the work cut out for them as manageable as possible. They even thought it fitting that all trains should run at the same (slow) speed in order to make short-term planning easier.

With hindsight, civilian railway experts criticized this approach and historians tend to agree with them. Sloppiness might have played a role, and it is possible that the Railway Office attracted less talent than the more prestigious Operations Office. Most striking is the lack of control over railway transport by the chief of the General Staff or the *Operationsbüro*. Probably, the subject matter was too complicated and boring to get more attention.[10]

With regard to 1914, the plans for two war cases are of particular interest: war case "R+B" and war case "B." In a war against Serbia (and perhaps Montenegro as well) without Russia's intervention, B-Group would fight on the Balkan Peninsula. Belgrade, Serbia's capital on the Danube, and the lower Moravia valley were situated along the Hungarian border. Further to the West, Bosnia-Herzegovina was separated from northwestern Serbia by mountain ranges and the Drina River. An offensive across the Danube and the Save rivers seemed appropriate. High-ranking officers such as the commander of the War Academy, Alfred Krauß, favored this approach, but Conrad would have none of it. The main thrust would be an eastward operation from Bosnia into Serbia, which would pin down Serbian troops in the northwestern corner of the country. Conrad had to keep in mind that B-Group might have to be redeployed at the beginning of the campaign. But Minimal Group Balkan alone could do little more than provide for an active defense of Bosnia-Herzegovina. Only by focusing on northwestern Serbia would a flexible response to a changing strategic situation be possible. Political considerations trumped geography.[11]

Prior to 1904, Russia had been far stronger than Austria-Hungary, but even after the war in Manchuria and the disruption of its military

[10] Norman Stone, "Die Mobilmachung der österreichisch-ungarischen Armee 1914," *Militärgeschichtliche Miteilungen* 16 (1974): 82–4; Degreif, *Operative Planungen*, pp. 221–3; Emil Ratzenhofer, "Die österreichisch-ungarischen Aufmärsche. Friedenspläne–Durchführung," *Schweizerische Monatsschrift für Offiziere aller Waffen* (1931), pp. 31–8, 73–7, 103–8; Bruno Enderes, "Die österreichischen Eisenbahnen," in Enderes, Emil Ratzenhofer, Paul Höger, eds., *Verkehrswesen im Kriege* (Vienna and New Haven, 1931), pp. 57–65.

[11] Rudolf Jeřábek, *Potiorek. General im Schatten von Sarajevo* (Graz, Vienna and Cologne, 1991), pp. 97–106.

by defeat and revolution, the tsar's army was the biggest headache of military planners in Vienna. Its huge territory and its seemingly unlimited manpower singled Russia out as an opponent which the Habsburg army would not be able to defeat. On the other hand, the Dual Alliance of 1879 made sure that Austria-Hungary could rely on German assistance in case of an unprovoked Russian attack. Since the wars of German unification, the Reich was perceived to be the strongest military power on the Continent, with reliable, well-trained troops, a first-class officer corps, and the most professional General Staff in Europe. Berlin, where the Prussian Ministry of War and the General Staff resided, was duly nicknamed the "Mecca of militarism." With such a strong ally by its side, Austria-Hungary would stand a fair chance to blunt any Russian attack. But there was a catch: since the 1890s, Germany also had to prepare for a two-front war, this against France and its new ally, Russia, which made it unlikely that the tsar's forces would have to face the entire German army at the beginning of a European war.

Russia's military resources, the speed of its mobilization, and the probable deployment of its troops had to be the basis for Austria-Hungary's war plan. From the 1880s to 1914, the General Staff in Vienna pinned its hopes on the higher speed of the k. u. k. army's mobilization and concentration. This implied that Austria-Hungary would not evacuate Galicia and the Bukovina. These Austrian crown lands bordering Russia lay to the east and to the north of the Carpathian Mountains, a natural barrier to invasion of the Habsburg Monarchy's heartland. Because of geography, eastern Galicia, the base for any forward deployment of Austrian troops, was rather isolated from the railway system. The General Staff made sure that railway connections with the rest of the Monarchy were improved in the 1880s and 1890s.

The concentration of the k. u. k. army in the area around Lemberg (Lvov, Lviv) offered some flexibility to operational planners. They could focus on an offensive between the Vistula and the Bug rivers, or on an attack in the direction of Dubno and Rovno to the Northeast. But the army's flank would be exposed to Russian assaults. Therefore, a major northeastern thrust had to be covered by an initial northward offensive or vice versa.

It was Conrad who decided that an initial northward thrust would be the better option. As Russia with French finances improved its European railway system substantively, an early offensive between the Vistula and the Bug looked less dangerous than the northeastern thrust favored by the *Operationsbüro*. It could be launched before B-Group would be concentrated in Galicia. After a victorious northward attack, Conrad planned for an eastward turn of the armies on the left wing, joining the right wing and forming a closed front facing the Russians to the North-east. An envelopment of these Russian forces would be the ultimate goal of the campaign.

To achieve at least a limited success in the initial operations, Conrad needed German support. In 1882 the chiefs of the General Staffs in Vienna and Berlin had deliberated on a war plan against Russia for the first time. The Germans promised to deploy more than half of their troops in the East; both allies agreed to launch offensives against the Russians in central Poland from eastern Galicia and East Prussia, respectively.

When Alfred von Schlieffen took over in Berlin as chief of the General Staff in 1891, he proposed a different tack. He wanted to attack from Posen into western Poland, shoulder-to-shoulder with the Austrian army. But the slowness of the k. u. k. army's mobilization and a lack of trust in its striking power made Schlieffen think twice. By 1897, both sides agreed to return to the old concept. During the Bosnian Annexation Crisis of 1908–09, Conrad seized the opportunity to seek an understanding with his new opposite number in Berlin, Helmuth von Moltke, the Younger. With the approval of the Foreign Ministry, Conrad started an exchange of letters with Moltke to inform each other about the basic strategic and – up to a point – the operational ideas for a war against Russia. Conrad learned about the most important aspect of Germany's war plan in case of a two-front war: the decision to focus on France first, leaving about 13 divisions in the East. After 40 days, the French army should have been defeated and the bulk of the German forces transported to the East. Austria-Hungary would have to bear the brunt of a Russian attack at the beginning of war.

Conrad accepted this burden-sharing without reserve. His own ideas for a two-front war echoed the strategic approach of both Schlieffen and Moltke. From a professional point of view, the German plan was simply the best solution. Unfortunately, the arrangements with Moltke left the k. u. k. army with the unrewarding task of having to defend the eastern parts of Germany and the Habsburg Monarchy – without enough troops left to attack Serbia, not to mention Italy. But in the most likely and at the same time most dangerous case, that Austria-Hungary was already at war with Serbia or Italy when Russia intervened, B-Group might make a difference before being redeployed. And offensive operations by the German Eighth Army, concentrated in East Prussia, would make it easier for A-Group to thrust northward – Conrad's favorite concept for the initial offensive.

Moltke had to make sure that Austrian forces would launch an offensive right at the beginning of a war because this seemed to be the only way to protect the eastern parts of Germany during the first six weeks. Conrad, for his part, promised to do so, but only if the Germans were attacking the Russians in the Narew area.

This *do-ut-des* ("I-give-so-that-you-may-give") arrangement formed the basis of Austro-German cooperation. In the following years, the General Staffs in Berlin and Vienna promised each other more than they could deliver because both sides depended on their respective partner's willingness to launch offensive operations in the East. As the feasibility of their war plans rested with these agreements, Conrad and Moltke fostered not only illusions but also deceived themselves. This is probably one of the reasons why the Austrians never really questioned the German plan to beat France within weeks and why the Germans shied away from scrutinizing the striking power of the k. u. k. army. This lack of a critical assessment of one's ally was not just a question of politeness. It was essential to believe in the alliance because otherwise there would not have been any way out of the strategic impasse.

Self-deception was made easier by the form of cooperation envisaged by Moltke and Conrad. With the Germans operating from East Prussia and the Austrians from eastern Galicia, there would be no need for a unified command of the coalition forces in the East. Nor was there any need to agree on specific plans or to train officers for joint operations. Later in the war, politicians and officers would criticize the narrow limits of prewar cooperation, but this was unfair. Given the standards of pre-1914 international relations and the strategic and operational plans of both allies, the General Staffs worked together quite well.

Conrad and Moltke would have been prone to self-deception in any case. A look at the agreements with Italy and Germany demonstrates the ambivalent effect of inter-allied cooperation. In order to bring Italian troops to Germany's western front in case of a Triple Alliance war, the railway experts of the General Staffs in Berlin, Rome, and Vienna had produced detailed transport plans. In the last two years of peace, there were new naval and transport agreements. The frenetic military diplomacy initiated by Moltke nourished hopes that Italy would honor its treaty obligations – not just in Berlin but also in Vienna. Without Italian support or at least Italian neutrality, a Great Power war was no longer an option. Conrad never really trusted Italy, but he no longer believed in an imminent Italian attack on Austria-Hungary. And in early 1914 even the deployment of Italian troops in Galicia was discussed in Vienna. It now is clear that Moltke and Conrad based their judgment on the balance of military power in Europe on an amateurish political calculation rather than on a professional assessment of troop numbers.

At the same time, the Viennese General Staff gave its up-dated estimate of the strategic situation in Europe. Approximately 130 divisions, 92 of them Russian, 16½ Serbian, 5 Montenegrin, and 16½ Romanian, would face 48 Austrian and 13 to 14 German divisions. It is not surprising that Romania's side-switching in 1913–14 made the Austrians nervous. But even if Romania had joined Germany and the Habsburg Monarchy, their foes would still have enjoyed a clear superiority in numbers. With regard to Russia's successful extension of its strategic railways and the higher speed of its mobilization, the *Operationsbüro* worked on plans for deployment of Austrian troops along the San and the Dnjestr. But Conrad rejected the proposal. It was essential to seize the initiative and offensive operations required a deployment closer to the border. Once again, Conrad confirmed that a northward thrust between the Vistula and the Bug offered the much-needed protection against Russian assaults on the flank of the k. u. k. forces.[12]

Austria-Hungary's most important war plan in 1914, the preparation for war case "R" in connection with war case "B," bore witness to the General Staff's unwavering bias toward the offensive. Conrad had advocated tactical innovations to let the offense prevail on the battlefield. He applied his favorite tactical maneuver, the out-flanking and eventual envelopment of the enemy, to operational tasks. Blinded by his enthusiasm for pincer-like offensive operations in Poland or Serbia, Conrad was willing to take high risks and to overtax the resilience of his troops.[13] He propagated preventive war as the miracle cure for Austria-Hungary's misery. Inevitably, the chief of the General Staff recommended offensive strategies too. No matter what kind of a task Conrad had to deal with, the recipe for success was to seize and keep the initiative. To him, offensive plans were the obvious solution to tactical, operational, or strategic problems.

[12] ÖStA, KA, Gstb, OpB 685, 739, 813; Conrad, *Aus meiner Dienstzeit*, vol. 3, pp. 529–32, 607–9. For the cooperation between the general staffs in Berlin and Vienna, see Kronenbitter, "*Krieg im Frieden*," pp. 277–314, which is partly based on the papers of the German military attaché in Vienna. Unlike the files of all the other German military attachés, these survived World War II by chance and include, among other things, hitherto unpublished correspondence on the July crisis between the military attaché and the General Staff in Berlin. See Günther Kronenbitter, "Die Macht der Illusionen. Julikrise und Kriegsausbruch 1914 aus der Sicht des deutschen Militärattachés in Wien," *Militärgeschichtliche Mitteilungen* 57 (1998), pp. 519–50.

[13] Hew Strachan, *The First World War*, vol. 1: *To Arms* (Oxford and New York, 2001), pp. 285–90.

Fit for Survival? War Preparations

Conrad inherited a General Staff that had seen a great deal of piecemeal improvement under Beck's command. The new chief of the General Staff spent a lot of energy on political quarrels and on ingenious operational concepts. Nevertheless, Conrad also paid attention to the more mundane aspects of his portfolio. With remarkable persistence, he strived for a root-and-branch reform of Austrian war preparations. Everything had to be geared to the needs of modern warfare: the selection of commanders, the training of officers and troops, and the equipment of the army. Conrad worked hard to instill the spirit of the offensive into the General Staff, field commanders, the officer corps, and the soldiers. Traditions had to be judged accordingly, and discarded if necessary.

This radical approach caused opposition from different quarters. Archduke Franz Ferdinand, for example, questioned the new training and field regulations. The heir to the throne believed in drill and discipline, not in the soldiers' capacity to judge the situation on the battlefield properly. Socialist politicians and conservative officers disliked the new form of maneuvers introduced by Conrad in order to get as close as possible to the harsh realities of modern warfare. Long marches and no official breaks during nighttime exhausted soldiers and horses. Nevertheless, most of the officers in the General Staff were proud of these "free" maneuvers because this innovation put the k. u. k. army ahead even of the Germans.

Maneuvers, like the staff rides, became physically very challenging and offered an opportunity to get rid of older commanders who had a hard time competing with their younger comrades. As war preparations became more rigorous and allegedly more realistic, the single-minded preference for the offensive dominated the thinking of the General Staff and the expectations of its officers. Maneuvers, staff rides, and war games served several purposes at the same time. On the one hand, operational doctrines and tactical regulations were put into "practice." On the other hand, officers were judged by their performance. Little wonder that only very few questioned the received wisdom of the General Staff. To expect that this kind of war preparation could put regulations, doctrines, and plans to the test would be asking for too much.[14]

To favor the offensive as the best solution to almost all tactical, operational and even strategic problems was in vogue. Not just in Austria-Hungary but in most of the Great Powers, the superiority of the offense

[14] Kronenbitter, "*Krieg im Frieden*," pp. 79–120.

dominated the experts' discourse. What made the k. u. k. army different was the discrepancy between the available resources and the ambitious operational plans. Since the beginning of the nineteenth century, the Monarchy had suffered from an inadequate fiscal system. In the wake of her defeat against Napoleon in 1809, Austria two years later went bankrupt and the financial situation proved to be unstable for most of the following decades. Losing a major war was costly in terms of money and prestige, but avoiding one in case of a clash of vital interests was not perceived as an option for one of Europe's Great Powers.

After 1867, financial constraints on the military build-up kept the army short of recruits. A modest increase in their number took place in 1889. Since the 1890s, the common army fell victim to the struggle between the Hungarian Diet and political leaders in Vienna. With so many checks built into the constitutional framework, Austria-Hungary lagged behind many of its potential friends and foes.

Raising the numbers of recruits in the reserves (Austrian Landwehr and Hungarian Honvéd) and improving their equipment was only a stop-gap measure. Sooner or later, Landwehr and Honvéd units would become integral parts of the first-line field army. The trademark of the k. u. k. army in the early phases of the war, the non-existence of a reserve army, was partly due to this *Ersatz* armament. The formation of new units for special branches like mountain artillery or machine guns came at the expense of the infantry. As a result, troop numbers in peace time were low and had to be increased by mobilization measures in case of a serious crisis.

Financial restrictions, bureaucratic infighting, and the conflict between Hungarian politicians and the Viennese elite had hampered army reform for years, until 1912 when István Tisza, the leader of the ruling party in Budapest, piloted a reform bill through the Legislature. The outbreak of the First Balkan War forced politicians to rethink budgetary priorities. From a relatively backward starting position, Austria-Hungary's land forces joined the armaments race of Europe's Great Powers. But even in 1913, the Habsburg Monarchy spent only slightly more on defense than Italy and about one-third of Russia's or Germany's defense expenditure. Whereas Russia and Italy devoted 5.1 percent of their net national product to defense, Austria-Hungary tried to get by on a defense burden of just 3.5 percent.[15]

[15] For the armaments race on the eve of World War I, see David G. Herrmann, *The Arming of Europe and the Making of the First World War* (Princeton, New Jersey, 1996);

Given the restrictions on discretionary spending, the priorities of Austria-Hungary's armaments program are quite telling. The build-up of the battle fleet swallowed an increasing share of overall military expenses. Franz Ferdinand was an ardent follower of "navalism." Like Wilhelm II, he loved the battle fleet as a symbol of assertiveness. Battleships were popular among industrialists too because they offered major building contracts; and to some segments of the broader public because they were symbols of technological progress. The army leadership accepted the costly dreadnought construction program until 1911, when a rise in the number of recruits was discussed seriously.

In 1913, War Minister Alexander von Krobatin and Conrad tried to curb the increase in the navy's budget, but they had a hard time fighting against the well-geared lobbyism of the Navy Command and Franz Ferdinand's bias. Navalism had an impact on Conrad and his circle just as it affected most of their contemporaries. This attitude and the hope that the money spent on the navy would either bring Italy to heel or help to defeat it, can explain the remarkable lack of inter-service rivalry. An investment in the battle fleet was better than no big armaments projects at all. Harder to understand was the sizeable amount of money for the construction of new fortifications along the Italian border. Also amazing was the fact that instead of modern quick-firing field guns, a rapid increase in the number of machine guns, or new infantry weapons, the k. u. k. army focused very much on mountain artillery and most of all on new heavy guns.[16]

What seems odd in the light of World War I makes perfect sense with regard to the political and strategic priorities of Conrad and the General Staff: to wage war on Italy ranked high on their agenda, followed by war plan "B." This looks like an irrational choice, given the unwillingness of the political leaders to attack the alliance partner and the probability of

David Stevenson, *Armaments and the Coming of War: Europe, 1904–1914* (Oxford, 1996). For figures on defense expenditure and on real defense burdens, see Stevenson, *Armaments*, pp. 4–6.

[16] For armaments, industry, and lobbying, see Kronenbitter, "*Krieg im Frieden*," pp. 179–96; Martin Gutsjahr, "Rüstungsunternehmen Österreich-Ungarns vor und im Ersten Weltkrieg: Die Entwicklung dargestellt an den Firmen Skoda, Steyr, Austro-Daimler und Lohner," PhD thesis (Wien, 1995); Manfred Reinschedl, *Die Aufrüstung der Habsburgermonarchie von 1880 bis 1914 im internationalen Vergleich: Der Anteil Österreich-Ungarns am Wettrüsten vor dem Ersten Weltkrieg* (Frankfurt am Main, 2001); Lawrence Sondhaus, *The Naval Policy of Austria-Hungary, 1867–1918. Navalism, Industrial Development and the Politics of Dualism* (West Lafayette, Indiana, 1994); Milan N. Vego, *Austro-Hungarian Naval Policy, 1904–14* (London and Portland, Oregon, 1996).

a Russian intervention in an Austro-Serbian war. Wishful thinking was involved, but there was more to the army's apparently odd allocation of funds: according to the General Staff's received wisdom, a two-front war against Russia and Italy or Serbia had to be fought as a sequence of two one-front wars. And because of Germany's strategic priorities (France), the strategic situation, and the Habsburg Monarchy's limited military resources, Italy or Serbia had to be attacked and defeated first. Heavy guns and mountain artillery were deemed essential to operations in the southwestern and southeastern theaters of war. Because of an argument about the tactical relevance of mobility and range, modern pieces were still in short supply when mountain artillery was deployed in 1914. But new heavy guns, replete with state-of-the-art truck transport, testified to the performance of Austria-Hungary's military-industrial complex – in case the General Staff pushed on with the project.

Airplanes, machine guns, and even field artillery did not get the same kind of attention. The shortcomings of Austria-Hungary's field artillery were well known, but only Schemua and Auffenberg questioned the striking power of the army in 1912. This was meant to win the politicians over to a budget increase and should not be mistaken as a profound criticism of the armaments program.[17] In the wake of the Balkan wars, the k. u. k. army could start to gear up for modern warfare.

Obviously, a major European war would be extremely costly. Financial preparations were an important aspect of war planning. A special credit of 2,556 Million Kronen (up from 2,517 in 1913) was designed to help pay for the costs of mobilization. Of this amount, 911 million was earmarked for the first 15 days of mobilization. Partial mobilization would be approximately 50 percent less expensive. Apart from the mobilization credit, Austria-Hungary was not very well prepared to finance a modern war. There was a rough idea of the costs of three months of war. But neither a long-term financial plan nor a war chest existed. In summer 1914 the Ministry of War called for immediate action – too late, as we now know.[18]

The same holds true for economic planning. In 1912 private entrepreneurs proposed precautionary measures on food supply for major cities in case of a war. The government responded to this initiative by gathering statistical data deemed necessary for successful planning, and by working out guidelines and procedures for the administration of food

[17] ÖStA, HHStA, PA XL 310; ÖStA, KA, MKSM 1912 25–1/6.
[18] ÖStA, FA, GFM 1914 267.

supply in wartime. From the military's point of view, it seemed to be even more important to maintain the Habsburg Monarchy's industrial capacities. According to Conrad, shortages of raw materials were one of the biggest problems. In July 1914, he urged an encompassing plan for Austria-Hungary's "economic mobilization" on the government. The Monarchy, clearly, was still without proper financial and economic plans when war broke out.

But in comparison to the state of war preparations prior to 1912, the improvements were quite remarkable. Both civilian initiatives and military planning had contributed to these changes. A case in point is the legislation regulating the state of emergency. The suspension of civil rights, including property rights, empowered the governments and the military commanders to suppress any kind of social or political disorder. Fear of the destructive consequences of a modern war and of nationalist or, to a lesser degree, socialist uprisings motivated the civilian elite.[19]

Officers in the General Staff and the Ministry of War began to react to sweeping changes in warfare. However reluctantly, they started to take into account that a modern war among Europe's Great Powers would have dramatic socioeconomic effects. There were signs of a new, broader concept of war preparations. But tradition, training and theory kept General Staff officers still focused on the mobilization and deployment of troops, and on operational plans. Modernity had its limits and its inconsistencies.

The cult of will power in an age of mushrooming bureaucracies and constraints on the individual is a prime example. The cult had its faithful believers in most European armies. In Austria-Hungary officers and civilians were afraid of the Monarchy's collapse in case of a Russian, Serbian, or Italian invasion of its border provinces. Troop morale was of great concern ever since the 1908–09 Annexation Crisis, when some rather isolated cases of mutinous behavior of Czech regiments occurred. Unable to use nationalism as a propaganda tool, the officer corps of the multiethnic army had to rely on martial tradition and social Darwinism to bolster its self-confidence. But basically, Conrad and his fellow officers were simply following the European trend. To seize and to keep the initiative came to be seen as a way to make operations, at least at the beginning of a war, more predictable. Little wonder that operation planners liked

[19] ÖStA, AVA, MRP Pressleitung 1914 526; ÖStA, KA, LVM Präs. 1913 3267; ÖStA, KA, Gstb 1913 62–14; ÖStA, KA, Gstb 1914 63–11; Kronenbitter, "*Krieg im Frieden*," pp. 227–32.

the idea. Because it fit in seamlessly, the bias toward the offensive went almost unchallenged.

Austria-Hungary Goes to War

When put to the test in 1914, Austria-Hungary's war plans – like those of the other Great Powers – proved inadequate.[20] The specific traits of Conrad's and the Viennese General Staff's failure had more to do with political considerations than with purely military affairs. When Conrad learned of Franz Ferdinand's assassination, he again called for war, but doubted that the Dual Monarchy would muster the resolve to attack Serbia. The peaceful solution to the crises in 1909, 1912, and 1913 had traumatized the chief of the General Staff and many others within the military elite. From their point of view, the worst outcome of a new confrontation with Austria-Hungary's unruly neighbor would have been another mobilization without war. According to them, this would have ruined the Monarchy's finances and the morale of officers, soldiers and the loyalists among the South Slavs. Without a victorious campaign against Serbia, the Habsburg Empire would fall prey to greedy neighbors and quarreling nationalities.

Until mid-July, Conrad was not sure whether this time would be different from the crises before. When he realized that Foreign Minister Berchtold would rather risk a Great Power war than to let Serbia get away with its support for ultranationalist and terrorist groups operating in Bosnia, Conrad became remarkably subdued – at least by comparison with his usual warmongering self. It was no longer necessary to harangue dovish politicians or a reluctant monarch. Now he and the General Staff would have to implement their plans.

Anxious to avoid a chaotic beginning to the war, the chief of the General Staff ruled that there would be no premature offensive against Serbia. Half the army was mobilized, but it took time to deploy it. This caused some concern among politicians in Budapest, Vienna, and Berlin.

[20] For Austria-Hungary's first campaigns, see Holger H. Herwig, *The First World War: Germany and Austria-Hungary, 1914–1918* (London, New York, Sydney, and Auckland, 1997), pp. 87–113; Jeřábek, *Potiorek*, pp. 118–49; Max Freiherr von Pitreich, *1914. Die militärischen Probleme unseres Kriegsbeginnes. Ideen, Gründe und Zusammenhänge* (Vienna, 1934); Manfred Rauchensteiner, *Der Tod des Doppeladlers. Österreich-Ungarn und der Erste Weltkrieg* (Graz, Vienna and Cologne, 1993), pp. 113–36; Norman Stone, *The Eastern Front 1914–1917* (London, 1998), pp. 70–91; and Strachan, *The First World War*, vol. 1, pp. 335–57.

But with regard to a possible Russian intervention, Conrad did not play it safe. Early in July he ordered the *Operationsbüro* to prepare for the deployment of A-Group in Galicia along the San and Dnjestr. This rearward deployment had been discussed before, under General Schemua and in March 1914, but had finally been rejected. Now, in July, it was meant to delay the final decision on a war against the tsar's empire.

At the same time, Conrad decided to mobilize one additional corps (III) for the offensive against Serbia. Second Army was to take part in the attack on the Serbs, and hence it was mobilized and deployed accordingly. In the last days of July it became obvious that Russia would intervene sooner rather than later, and Austria-Hungary switched from partial to general mobilization. Nevertheless, Conrad postponed the decision to redeploy Second Army in Galicia. After the war, the chief of the General Staff passed the buck to the Railway Office because the experts there had objected to an earlier redeployment. But a closer look at the evidence clearly shows that it was Conrad who wanted to play for time. If there was any logic in his decision, he was probably still hoping that the Germans might force the Russians to back down, or at least to enter the fray with some delay. Unlike 1912–13, it would be up to the German army to call Russia's bluff. In this, Conrad succeeded.

In terms of military effectiveness, the belated relocation of Second Army and the rearward deployment in Galicia were damaging. The importance of both decisions should not be exaggerated, but there is no denying that they made an already difficult mission even more challenging. Neither the General Staff nor the army was up to this task. Second Army could not stay in the Southeast long enough to make a difference there. The attack on Serbia failed. In Galicia, the first operations were quite successful, but the failed Battle of Lemberg sealed the fate of Austria-Hungary's offensive.

Conrad tried to blame it on the Germans because they had not come to the k. u. k. army's help by launching an offensive against the Russians in the Narew area. To be sure, the timing and direction of German operations aggravated the already difficult situation in Galicia, but this critique was wide off the mark and Conrad probably knew it. Germany's Eighth Army defeated and tied up more Russian troops than could have been expected by any General Staff officer worth his salt. The chaotic deployment of troops, bad reconnaissance and leadership, tactical inflexibility, and a lack of modern field guns were responsible for the defeat in Galicia. Unable to stick to his plan, Conrad did not part with his doctrine of the superiority of the offense. The yawning gap between ambitious plans and limited resources was bad enough; Conrad's miscalculations in summer 1914 made it worse.

The new professionalism and vigor in Austria-Hungary's war planning since 1906 were not accompanied by military or political oversight of the General Staff's work. All over Europe, modern, more professionalized forms of leadership and a lack of inter-bureaucratic coordination and supervision went hand in hand. Moreover, hierarchy and prestige left little room for internal debate about alternatives to Conrad's ideas. Unfortunately, the chief of the General Staff did not pay enough attention to logistics and the limited physical resilience of the troops.

In reaction to a foreign policy that tried to defend Great Power status on the cheap, military leaders in Vienna, just as their opposite numbers in Berlin, became obsessed with strategic speculation. The question whether and when Romania or Italy would switch sides in the future trumped the thorough analysis of war readiness. In 1914, Conrad focused too much on politics and neglected the less glamorous parts of his craft.

3

German War Plans

Annika Mombauer

Imperial Germany, unified just over forty years before the outbreak of World War I, was the result of war – Prussia's victories in the wars of 1864, 1866, and 1870–71. It is tempting, of course, to view the history of Imperial Germany as one that inevitably led to the war of 1914, given such violent beginnings and given that we know how it ended, but there was nothing preordained or inevitable about the German Empire's demise. Rather, successive political and military leaders were responsible for Germany's increasing international isolation. At the heart of Europe and surrounded by potential enemies, its fate was, arguably, sealed when the cautious diplomacy of Otto von Bismarck was abandoned in favor of Kaiser Wilhelm II's impulsive and willful foreign policy.

Coupled with this was a pronounced influence of certain military leaders and indeed of all things military. Germany's political leaders exerted little or no control over military decision-makers and war planning was conducted by them almost in a vacuum. Convinced of Germany's increasing "encirclement," the country's successive chiefs of the General Staff had the unenviable task of trying to prepare the German army for a future war in which the country might well be surrounded by hostile neighbors whose military manpower far outweighed that of Germany. Put differently, the General Staff's planning had to reflect the diplomatic context of the European system as well as the power potential of likely future opponents. As France and Russia concluded their alliance by 1894, it became clear that Germany would have to avoid a two-front war if possible.

A solution had to be found to escape Germany's geographic and strategic dilemma in case of war. Under the Elder Helmuth von Moltke, the victor of the "wars of unification," the solution had been to concentrate on the East in any future deployment. But under his successors, the focus of German war planning changed to the West. With Russia temporarily weakened by the Russo-Japanese War and the Revolution of 1905 and slow to mobilize, a chance presented itself to concentrate on France. Imperial Germany's war plan in 1914 was essentially to avoid a war on two fronts by turning it into successive one-front wars. This plan is commonly known by the name of the chief of staff who initiated the shift from east to west, Count Alfred von Schlieffen.

As the editors of this volume suggest, most educated citizens are likely to know "the basics" of Germany's war plan, whereas only a few specialists would know anything at all about the war plans of the other great powers.[1] Although failing to deliver a German victory, its "success" lay in becoming the stuff of legends and being elevated to an alleged recipe for victory. After the war, a mythical narrative was deliberately constructed by those who had been involved in Germany's military planning. The widespread view of the Schlieffen Plan as a panacea for victory for encircled Germany is the result of the historical writing of a "Schlieffen school" of military historians who aimed to prove after the war that Germany could have won with more capable leaders.

Underlying this myth are several arguments advanced by supporters of Schlieffen. They argued that the general should not have been replaced by the Younger Helmuth von Moltke in 1905 and that the Schlieffen Plan could have delivered the much-wanted speedy victory in the West and ultimately a general victory for Germany had it not been adulterated by Moltke.[2] Certainly, the mythmakers were successful in ensuring that over a hundred years after Schlieffen retired, his military plan is still legendary and that the "basics" of it are still known today, even outside military circles and even outside Germany. The same can certainly not be said of France's "Plan XVII," Russia's "Plan 19," or Austria-Hungary's Plans "R," "B," and "I."

Germany's war plan differs in another way when compared to those of its neighbors. Other war plans have not usually been seen as the cause of World War I, whereas in the German case, the Schlieffen Plan has become inexorably linked to the war-guilt question and has often been seen to be a causal factor in the events that led to the outbreak of war.

[1] See Ch. 1 above, p. 3. The Schlieffen Plan figured prominently in a popular history by Barbara W. Tuchman, *The Guns of August* (New York, 1962).

[2] On the Schlieffen myth, see Annika Mombauer, *Helmuth von Moltke and the Origins of the First World War* (Cambridge, 2001), particularly chapter 2.

Of course, every country had war plans, and they tended to be offensive. The existence of the Schlieffen Plan (or rather, in 1914, the Moltke Plan), in itself is certainly no proof of war guilt. But the other major powers did not implement their plans until after Austria-Hungary's attack on Serbia and Germany's offensive in the West initiated hostilities. Germany's action, it is often said, forced reactions by the military and political leaders of Europe's major powers, leading them to put their own war plans into action. Little wonder, then, that to many contemporaries (and later to many historians), Germany appeared guilty of starting the war of 1914.

There is a further important difference between Germany's war plan and those of its enemies and ally. The nature, even the very existence, of that plan has been contested in a recent heated historiographical debate. Historians seem to be interested in the Schlieffen Plan, in its minutest detail, in a way that had not been shown for the plans of the other European powers. In recent years, they have been forced to revisit the Schlieffen Plan following claims that the plan never existed. Moreover, it has been maintained that German strategy in 1914, far from being offensive in nature, was in fact a defensive plan designed to deal with a Franco-Russian attack. In Terence Zuber's opinion, "civilians, academics and Anglo-Saxon officers" have for too long believed that the Schlieffen Plan existed and that this "was the way great strategy should look.... The concept of the Schlieffen plan was so simple that every commentator felt he understood it and was justified in passing an opinion on it."[3]

This chapter will introduce the planners, examine the perceived threats of the prewar period, characterize the war plan, and review both its implementation and outcomes. It will also engage (albeit briefly) the recent debate around the German war plan.[4] In addition, it needs to consider two war plans: the "infamous" Schlieffen Plan, and the war plan of his

[3] Terence Zuber, *Inventing the Schlieffen Plan: German War Planning 1871–1914* (Oxford and New York, 2003), p. 34. In his most recent contribution to the debate, he is even more damning of those who believe that there was a Schlieffen Plan. Zuber, "Der Mythos vom Schlieffenplan," in Hans Ehlert, Michael Epkenhans and Gerhard P. Groß, eds., *Der Schlieffenplan. Analyse und Dokumente* (Paderborn, 2006), pp. 45–78; esp. pp. 45–46, 78.

[4] Too much ink has already been spilled as a result of this controversy. This chapter is not another contribution to the long and by now futile debate around Zuber's thesis. The debate can be considered exhausted, particularly following the publication of Ehlert et al., *Schlieffenplan*. My views on the debate have been made clear there: Annika Mombauer, "Der Moltkeplan. Modifikation des Schlieffenplans bei gleichen Zielen?," ibid., pp. 79–99; and in Annika Mombauer, "Of War Plans and War Guilt: The Debate Surrounding the Schlieffen Plan," *Journal of Strategic Studies* 28 (October 2005): 857–85.

successor. After all, Schlieffen retired from office in December 1905. For the next nine years, the Younger Moltke was in charge of military planning and adapted the nation's basic war plan to changing circumstances. Germany's war plan as of 1914 was no longer Schlieffen's. The operative plan should instead carry the name of the man responsible for German strategy on the eve of war, Moltke. In this investigation we need to be mindful that there were two different, if very similar and at times identical, war plans and two different war planners. The analysis of the perceived threats that Germany faced and the perceived needs of the armed forces to prepare to deal with them will demonstrate both similarities and differences in the way Schlieffen and Moltke viewed Germany's position vis-à-vis its neighbors.

The Planners and Their *Weltbild*

The problem of a war on two fronts had occupied Germany's first chief of the General Staff, the Elder von Moltke, ever since Prussia had defeated France in 1871. Following the experience of the Franco-Prussian War, Moltke doubted the possibility of a total victory against two opponents. In April 1871 he wrote: "Germany cannot hope to rid herself of one enemy by a quick offensive victory in the West in order then to turn against the other. We have just seen how difficult it is to bring even the victorious war against France to an end."[5] As the French began to fortify their eastern border, Moltke believed that these fortifications would make an offensive in the West impossible. As a result, in a future war he wanted to concentrate on an initial attack in the East, albeit with a limited objective. According to historian Gerhard Ritter, Moltke "was content with a defensive which was to exploit any opportunity for offensive thrust."[6]

During Alfred von Waldersee's brief period as Moltke's successor, few changes were made to this plan. Waldersee was replaced in 1891 by the 58-year-old Alfred von Schlieffen.[7] During his time as chief of the

[5] Cited in Gerhard Ritter, *The Schlieffen Plan: Critique of a Myth* (London, 1958), p. 18.

[6] Ibid., p. 21.

[7] Count Alfred von Schlieffen was born in Berlin on 28 February 1833 and died 4 January 1913. His army career began in 1853. He fought in the war against Austria in 1866 and in the Franco-Prussian War. He was chief of the German General Staff from 1891 until 1905. His reputation as a taciturn workaholic and military genius was partly created by his former colleagues and supporters who created the myth of Schlieffen as the harbinger of a secret recipe for victory. As Arden Bucholz points out: "It is strange that Schlieffen was so venerated for he was only the author of a plan that failed." Arden Bucholz, *Moltke, Schlieffen and Prussian War Planning* (Providence and Oxford, 1991), pp. 109ff.

General Staff from 1891 to 1905, Schlieffen altered the Elder Moltke's strategic plan. While Moltke had intended to split the German army into two roughly equal parts, Schlieffen wanted to amass most of it on one front, where a decisive battle of annihilation was to be fought. After the Russians, apparently due to their knowledge of Germany's strategic plan, started to concentrate their troops in their northern and western military districts and began to fortify places along the Niemen and Narew rivers and increased their defensive capabilities, Schlieffen began to doubt the feasibility of an offensive in the East.[8]

Instead, he saw Germany's best chance of victory in a swift offensive against France, while in the East the German army was initially to be on the defensive.[9] He planned to deal with Russia after France had been delivered a decisive blow. Schlieffen shared Moltke's concerns about the French fortifications; Germany's troops would have to avoid them as they might lead to position warfare of inestimable length. Instead, the opponent's armies should be enveloped.[10] Moving through Switzerland was not considered seriously as an option, not least due to geographic constraints and to the fact that the Swiss army was a force to be reckoned with, whereas north of France, Luxembourg had no army at all and the weak Belgian army was expected to retreat to its fortifications.[11]

[8] Wilhelm Dieckmann, "Der Schlieffenplan," unpublished manuscript, no date, Bundesarchiv-Militärarchiv (BA-MA), W10/50220, pp. 8–9.

[9] In his diary, Waldersee recorded that the impetus for changing the direction of the offensive from the East to the West came from Wilhelm II, rather than Schlieffen. Heinrich Otto Meisner, ed., *Denkwürdigkeiten des Generalfeldmarschalls Alfred Grafen von Waldersee* (3 vols. Stuttgart 1923–25), vol. II, p. 318. Dieckmann agrees ("Schlieffenplan," pp. 47–48), but where Waldersee blamed the kaiser's naïvety ("*unreife Ideen*"), he identifies political considerations behind Wilhelm II's wish to alter the strategic plan because at the time (1894), Wilhelm was eager to improve relations with Russia. A change of military strategy was thus in keeping with political aspirations.

[10] Schlieffen was inspired by the historic battle of Cannae of 216 B.C., at which the Carthaginian military leader Hannibal defeated the numerically superior Romans with one of the most famous envelopments in history. See Martin Samuels, "The Reality of Cannae," *Militärgeschichtliche Mitteilungen* 1/1990: 7–31. Yet crucially, Schlieffen failed to notice that although the Romans were defeated at Cannae, they won the overall battle on account of their superior sea power. See Holger H. Herwig, *The First World War: Germany and Austria-Hungary 1914–1918* (London, 1997), p. 49; Wolfgang von Groote, "Historische Vorbilder des Feldzuges 1914 im Westen," *Militärgeschichtliche Mitteilungen* 1/1990: 33–55.

[11] Dieckmann, "Schlieffenplan," pp. 121–22. On the question of Switzerland, see Hans Rudolf Fuhrer and Michael Olsansky, "Die 'Südumfassung.' Zur Rolle der Schweiz im Schlieffen- und Moltkeplan," in Ehlert et al., *Schlieffenplan*, pp. 311–38.

Schlieffen decided to concentrate all German effort on the right wing, even if the French opted for an offensive along another part of the long common border, thus taking the risk of allowing the French temporarily to reclaim Alsace-Lorraine.[12] In his last maneuver critique, Schlieffen outlined how it was possible for a smaller army to defeat a larger one. To avoid being swallowed up by the opponent in a frontal attack, it would have to advance in the most sensitive areas – the flanks and the back, and aim to catch the opponent by surprise.[13] While roughly 15 percent of German forces (along with the Italian Third Army) would secure the German left flank in Alsace-Lorraine and along the Upper Rhine River, the vast majority of troops would be concentrated on the right wing in order to march through Luxembourg, Belgium, and The Netherlands, "sweep" the English Channel with their right "sleeves," and then descend into the Seine basin southwest of Paris, where they would destroy the main French forces. In the meantime, a single German army would hold against the Russians in East Prussia. The campaign was predicated on speed: Paris had to fall and the French armies be destroyed by the 39th or 40th day of mobilization. Instead of a war on two fronts, two wars were to be fought in quick succession. If Germany was to have any chance of victory, the war had to be short.

In December 1905, the basic premise of Schlieffen's strategic thinking became a blueprint for his successor, the Younger Moltke.[14] Not only did the two men share the same ideas regarding Germany's war plan, but also similar fears of what the future held for Germany.[15] In memoranda and personal testimonies, both Schlieffen and Moltke have left us plenty of clues as to their mindset and their perception of threats that Germany faced in the early twentieth century. "Encirclement" was the

[12] BA-MA, Nachlass Tappen, N56/2, letter from Reichsarchiv to Tappen, December 1923, p. 255.

[13] Bayerisches Hauptstaatsarchiv-Kriegsarchiv (BHStA-KA), Bestand Generalstab 1237, Schlussbesprechung 23 December 1905, p. 13.

[14] Helmuth Johannes Ludwig von Moltke, nephew of the Elder Moltke, was born on 25 May 1848 in East Prussia and died 18 June 1916 in Berlin. In the Franco-Prussian War, he fought in the battles at Weißenburg, Wörth and Sedan, and participated in the siege of Paris. In 1880 he joined the General Staff, and in 1882 became his uncle's personal adjutant. After the Elder Moltke's death in April 1891, Wilhelm II made Moltke his personal aide-de-camp (*Flügeladjutant*). At the time of his appointment to head the General Staff, he had reached the rank of *Generalquartiermeister*, or deputy chief of staff. For Moltke's biography and military career, see Mombauer, *Moltke*, pp. 46–54; and Bucholz, *Prussian War Planning*, pp. 216ff.

[15] For a detailed discussion of the similarities and differences between the two men's war planning see Mombauer, "Der Moltkeplan," in Ehlert et al., *Schlieffenplan*, pp. 79–99.

most frightening of the predictions. Ever since Bismarck had worried about hostile alliances being formed against the recently unified German Reich, the idea of an encircled Germany haunted military and political planners. The nation's leaders saw themselves with one reliable ally, Austria-Hungary, whose military efficiency was doubted by many leading military men; and another, Italy, whose support the military planners desperately hoped for, despite their better judgment. At the same time, they would be surrounded by the armies of their hostile neighbors: France and Russia, possibly also Britain. Schlieffen described the threat with these words: "the gates will be opened, the drawbridges let down, and armies of millions of men will pour into Central Europe across the Vosges, the Meuse, the Königsau, the Niemen, the Bug, and even across the Isonzo and Tyrolean Alps in a wave of devastation and destruction."[16]

The military planners' views on likely friends and foes changed depending on international tensions or crises, but certain key concerns remained constant. There was the question of who the real enemy was: France or Russia? There was the fear of losing the support of Austria-Hungary and the increasing worry over the reliability of Italy. And there was the concern that for the ambitious war plan to succeed, Germany needed a lot more troops than the chiefs of staff had at their disposal. There was also a fear of losing the "opportunity" for a war altogether; this brought Moltke's demand in December 1912 for a war "now or never."[17] Time seemed to be running out for Germany's war plan as its likely enemies were increasing their armies and tightening their bonds with each other.

At various times, Moltke communicated his fears to German political leaders. On 2 December 1911, shortly after the Moroccan Crisis had been resolved, and against the background of the Italo-Turkish War over Tripolitania and Cyrenaica, he sent a memorandum to the chancellor highlighting some of his fears of what the future held.[18] With the imminent army bill in mind, Moltke speculated as to who the likely opponents

[16] Cited above, p. 19; from Alfred von Schlieffen, "Der Krieg der Gegenwart," BA-MA, Nachlass Schlieffen, N43, vol. 101.

[17] This was the so-called war council meeting. See John C. G. Röhl, *The Kaiser and his Court: Wilhelm II and the Government of Germany* (Cambridge, 1994), pp. 162–189; and Mombauer, *Moltke*, pp. 135ff.

[18] This memorandum was excluded from the official publication of documents in *Der Weltkrieg*, but was paraphrased in the text. Reichsarchiv, *Der Weltkrieg. Kriegsrüstung und Kriegswirtschaft, Anlagen zum Ersten Band* (Berlin, 1930), vol. I, pp. 11ff. It can be found in the unpublished *Reichsarchiv* study "Die Militärpolitische Lage Deutschlands," BA-MA, W-10/50276; and in Annika Mombauer, ed., *The Origins of the First World War in Europe: A Document Collection* (Manchester, forthcoming, 2010).

in a future war might be. France, he thought, probably did not want to go to war against Germany at that time, but there existed in France a chauvinistic mood that could provoke a war. If it came to war, he predicted, Britain would actively support France. Italy's support could not be relied upon – another accurate prediction – despite its contractual obligations as Italy's coasts were exposed to a British attack and as Italy's leaders lacked the desire to fight France. Similarly, Austria-Hungary was under no obligation to support Germany against France and Britain. Germany, Moltke thought, should prepare itself to face such a contingency on its own. If, however, it turned out that Russia was obliged to assist France, this would mean the *casus foederis* for Austria-Hungary. Turkey's role should also not be underestimated, were it to participate on the alliance's side. Turkey, Moltke argued, was the only power that could threaten Britain on land (Suez Canal, Egypt, Aden), an argument that he would advance in 1914. Moreover, in case of Turkish intervention, Russia would be forced to deploy troops in the Caucasus which could otherwise be used on the German-Austrian front.

The only certain factor, Moltke believed, was that Germany and Austria faced a coalition of France, Britain and Russia, whose military power had increased over the last few years. Russia was reorganizing and increasing its army, equipping it with better *matériel*, simplifying mobilization procedures, extending the strategic railway system, and providing for a younger officer corps. Moltke warned that it was no longer true – as it had been in 1905–06 – that Russia would not be able to wage a European war. Fear of Russia's increasing military potential now became a recurrent theme in Moltke's reasoning, but it was not the only threat Germany faced. Britain would wage war primarily at sea, Moltke opined rather strangely and without basis (as Keith Wilson shows in his chapter), but would also send a well-equipped army of 150,000 men to the Continent. And France, the memorandum concluded fatalistically, enlisted its entire population in such a way that Germany was left far behind.

> The equipping, perfection and strengthening of its military power in all areas makes France an ever more powerful and dangerous opponent. Everyone prepares themselves for the big war that is widely expected sooner or later. Only Germany and its ally Austria do not participate in these preparations.[19]

[19] "Die Militärpolitische Lage Deutschlands," BA-MA, W-10/50276, pp. 14–17; Reichsarchiv, *Weltkrieg*, vol. I, pp. 2–13. For a discussion of Moltke's memorandum, see David Stevenson, *Armaments and the Coming of War: Europe 1904–1914* (Oxford, 1996), pp. 202–3; and David G. Herrmann, *The Arming of Europe and the Making of the First World War* (Princeton, 1997), pp. 169ff.

This memorandum was intended to convince the chancellor and others of the need for further army increases. At the end, Moltke urged that both a further expansion of the fleet and an increase of the army's peacetime strength were "a prerequisite for self-preservation. Both have to go hand in hand."[20] Those fears were echoed by Moltke constantly in the years leading to the outbreak of war and were not solely motivated by his desire for army increases.

Schlieffen Plan and Moltke Plan: The Nature of Germany's War Plans

The basic premise of Germany's war plans in the years 1905–14, as indicated, was to focus initially on the weaker enemy, France, aiming for a speedy and decisive victory; and then turn to the enemy in the East, Russia, before that nation had time to gather its main forces. With hindsight one can see that both Schlieffen and Moltke underestimated the strength and determination of Germany's future enemies and thought that a numerical inferiority on Germany's side could be matched with superior morale and training.[21] They both feared, however, that the enemies would, in the near future, become too strong. Their daring war plans would not lead the Reich to victory indefinitely, which is why, at different important junctures, both men favored preventive war. During a state council meeting at the height of the First Moroccan Crisis in 1905, for example, Schlieffen claimed:

> Russia is tied up in the East, England still weakened by the Boer War, France still behind with its armaments. Before long the German Reich has to prove its worth through a war. Now is the most convenient time. Therefore my solution: war with France.[22]

[20] "Die Militärpolitische Lage Deutschlands," BA-MA, W-10/50276, p. 43.

[21] It was no secret that German troops were numerically inferior to their opponents. See "Welche Nachrichten besaß der deutsche Generalstab über Mobilmachung und Aufmarsch des französischen Heeres in den Jahren 1885–1914," BA-MA, W-10/50267 (RH61/398). For similarities and difference in Schlieffen's and Moltke's planning, see Robert T. Foley, "Preparing the German Army for the First World War: The Operational Ideas of Alfred von Schlieffen and Helmuth von Moltke the Younger," *War & Society* 22 (October 2004): 1–25.

[22] Wilhelm von Hahnke reporting on this meeting in a letter to Wilhelm Groener in 1926. Schriftwechsel betr. Schlieffen-Plan, 1919–1932, 16 April 1926, BA-MA, N46/38, pp. 49ff. On other occasions (1909, 1911, 1912, and 1914), Moltke, too, demanded a preventive war. Historians are divided on the question whether Schlieffen demanded preventive war in 1905. For further evidence indicating Schlieffen's and Moltke's desire to fight a preventive war see Mombauer, *Moltke*, pp. 44, 78, 104. For a different view, see Ehlert et al., *Schlieffenplan*, pp. 8–9.

Moltke, too, demanded war at various points, most notably, as noted, during the so-called war council of December 1912.[23]

We can trace the development of the German war plan with the help of deployment orders (*Aufmarschanweisungen*) which were drafted and updated annually. Copies of these have survived in the German military archive, and we now fortunately have them in published form.[24] These documents reveal how German strategic thinking shifted from a plan that concentrated on the East to one that favored a western deployment and, from 1913, to a plan which only allowed for an offensive in the West. The Eastern offensive had been abandoned. France was the main target for an initial offensive.

Moltke, like Schlieffen before him, wanted to fight two wars in succession, rather than a war on two fronts, and this required a quick victory against France in the West. He explained this to the Austro-Hungarian Chief of Staff, Franz Baron Conrad von Hötzendorf, in 1909:

> Our foremost intention must be to achieve a speedy decision. This will hardly be possible against Russia. The defense against France would absorb such great strength that the remaining troops for an offensive against Russia would not suffice to force a decision onto it.[25]

Moltke laid out the principles of his war planning in a memorandum of 2 December 1911 where he made it clear that defeating France had to be Germany's immediate priority.

> [T]he decision of the war is in the fight against France. The Republic is our most dangerous enemy, but we can hope to achieve an early decision here. Once France has been defeated in the first great battles, then the country, which does not have large reserves of people, will hardly be able to continue with a long war, while Russia could divert [the war] after a lost battle into the interior of its vast territory and drag it out for an inestimable time. Germany's entire striving must, however, be aimed at ending the war at least on one side as quickly as possible by way of a few great strikes.[26]

[23] See reference 17 above.

[24] They are contained in Ehlert et al., *Schlieffenplan*, pp. 341–484.

[25] Moltke to Conrad, 21 January 1909. "Forschungsarbeit gemeinsame Kriegsvorbereitung Deutschland und Österreich-Ungarn," BA-MA, N46/38 p. 15. That Moltke and his colleagues, just like Schlieffen before them, saw Germany's best chance for a speedy victory in the West, rather than the East, can also clearly be seen in the deployment plans of the prewar years. "Aufmarschanweisungen für die Jahre 1893/94 bis 1914/15," BA-MA, RH61/v. 96. Now published in Ehlert et al, *Schlieffenplan*, pp. 341–484.

[26] Memorandum of 2 December 1911. "Die militär-politische Lage Deutschlands," BA-MA, W-10/50279, p. 19.

Even the Kaiser, who had been kept in the dark about the details of military planning, was aware of the direction in which "his" troops would march if it came to war. During the Bosnian Crisis of 1909, he was worried that the *casus foederis* might arise for Germany if Austria-Hungary and Russia ended up going to war. He feared that this would "result in mobilization and the war on 2 fronts for Germany, i.e. in order to march on Moscow, Paris has to be taken first."[27]

In addition to the general direction of a future offensive, there was also agreement as to how the war in the West was to be conducted. Both Schlieffen and Moltke planned to avoid France's eastern fortifications in order to be able to deal it a decisive blow without getting bogged down in a long war. To do so, however, would mean moving through Luxembourg, Belgium (and for Schlieffen, The Netherlands). Doing so, of course, meant the violation of Belgian neutrality, a move that would do much to damage Germany's reputation.

The idea of enveloping the opponent by violating Belgian neutrality was not new. Following the "Boulanger crisis," Bismarck had inquired of Waldersee whether it would not be practicable to march through neutral Belgium.[28] Far from having been "Schlieffen's ingenious idea," a violation of neutral Belgium had been a frequently discussed topic, just as it was of serious concern to German strategists whether France might begin hostilities by marching through Belgium.[29] The first written record of Schlieffen's plan to violate Belgian neutrality dates from August 1897. As he put it bluntly: "An offensive which aims to circumvent Verdun must not shy away from violating not only the neutrality of Luxembourg but also that of Belgium."[30]

After 1905, Moltke was of the same opinion. In December 1912 he explained to the chancellor how an offensive against France and a defensive stance against Russia were to be conducted:

> [I]n order to be on the offensive against France it will be necessary to violate Belgian neutrality. Only by advancing through Belgian territory can we hope to attack the French army in the open [*im freien Felde*] and defeat it. This way we will face the English Expeditionary Corps and – if we do not manage to

[27] *Die Große Politik der Europäischen Kabinette 1871–1914. Sammlung der Diplomatischen Akten des Auswärtigen Amtes*, eds. Johannes Lepsius, Albrecht Mendelssohn-Bartholdy, and Friedrich Thimme (40 vols, Berlin, 1922–1927), vol. 33, No. 12349.

[28] Dieckmann, "Schlieffenplan," p. 57. Georges Boulanger became war minister in January 1886 and at once stirred national passions for a war of revenge against Germany.

[29] Ibid., pp. 60–63.

[30] Quoted in ibid., p. 64.

come to an agreement with Belgium – also Belgian troops. However, this operation is more promising than a frontal attack against the fortified French eastern front.[31]

For his strategy to succeed, Moltke explained, it was necessary to march through neutral Belgium and into France. It mattered little to either Moltke or Schlieffen whether France was planning to do the same (and it is worth noting that a possible French invasion of Belgium was never cited as an "excuse" by Schlieffen or Moltke for their plans). Regardless of the French war plan – which eluded German military intelligence – the majority of troops would be employed on the right wing in an effort to outflank the French and avoid the fortified border. This was the famous – even legendary – "right wing" of the German army, charged with attempting to outmarch and ultimately outflank the French. Both Schlieffen and Moltke agreed that this was the best, if not the only option, if Germany was to achieve a speedy victory.

However, because Moltke's planning differed from Schlieffen's in several important ways, significant changes were made to the German plans from 1909 onward. Moltke was often at pains to outline how his changes to Schlieffen's intentions resulted in a better strategy for Germany.

One of the most important differences resulted from Moltke's fears of the effects of an allied blockade on Germany (to be taken up later). Where Schlieffen had advocated marching the right wing armies through the neutral Netherlands as well as Belgium and Luxembourg, Moltke wanted, if possible, to ensure that The Netherlands would remain neutral and serve Germany as a "windpipe" in a future war.[32]

Changing the deployment plan in the West to ensure that neutrality was preserved created its own problems, not least because the massive armies of the right wing now had to march through a relatively narrow "corridor" into Belgium. It necessitated the well-known *coup de main* (bold stroke) on Liège, the strategically important fortress that had to be brought under German control if the advance of the right wing was to proceed as planned. This, in turn, required an almost immediate dispatch of troops into neutral Belgium and Luxembourg before war was even declared, as was indeed the case in August 1914. To overcome some of these difficulties, Moltke attempted, unsuccessfully, to come to an

[31] Moltke to Bethmann Hollweg, "Über die militärpolitische Lage," 21 December 1912. Reichsarchiv, *Der Weltkrieg*, No. 54, pp. 163–4.

[32] For a more detailed discussion of these differences in planning, see Mombauer, "Der Moltkeplan," in Ehlert et al., *Schlieffenplan*, pp. 79–99.

agreement with Belgium to allow German troops unhindered access to France via its territory, most famously during a meeting with King Albert at Potsdam in November 1913.[33]

Moltke wanted to avoid marching through the Dutch province of Limburg because he did not consider this militarily necessary.[34] In a marginal note on Schlieffen's 1905 memorandum, he had remarked that it would be of immense advantage to Germany if it were possible to come to an arrangement with The Hague by diplomatic means, because the German deployment plan would benefit from the Dutch railways and The Netherlands as an ally would be "of inestimable value."[35] Moltke in the first year of the war compared his plans directly with Schlieffen's. Thus, in November 1914 he wrote:

> Count Schlieffen even wanted to march with the right wing of the German Army through southern Holland. I changed this in order not to force The Netherlands also onto the side of our enemies, and preferred to take on me the great technical difficulties that were caused by the fact that the right wing of our army had to squeeze through the narrow space between Aachen and the southern border of the province of Limburg.[36]

He repeated this position in a letter of 1915. "In contrast" to Schlieffen, he noted, "I predicted that the German right army wing would lose such strong forces because of a hostile Holland that it would lose the necessary force [*Stoßkraft*] against the West."[37]

The "technical difficulties" that arose from this change of plan were extremely grave. Advancing the huge numbers of troops of First and Second Armies into Belgium was the most taxing problem. If the deployment plan was to succeed at all, Liège had to be taken, as Moltke noted looking back in 1915: "The fortification had to be in our possession if the advance of the 1st Army was to succeed at all. This realization led me to the decision to take Liège by *coup de main*."[38]

Eighty years later, Terence Zuber would deny that such a bold stroke and the march through Belgium toward France were ever Moltke's

[33] For details of the visit, see Jean Stengers, "Guillaume II et le Roi Albert à Potsdam en novembre 1913," *Bulletin de la Classe des Lettres et des Sciences Morales et Politiques* (1993): 7–12; Mombauer, *Moltke*, pp. 153–67.

[34] "Die Militärpolitische Lage Deutschlands," BA-MA, W-10/50276, pp. 71–72.

[35] Gerhard Ritter, *Der Schlieffenplan. Kritik eines Mythos* (Munich, 1956), p. 148.

[36] *Helmuth von Moltke, Erinnerungen, Briefe, Dokumente 1877–1916. Ein Bild vom Kriegsausbruch, erster Kriegsführung und Persönlichkeit des ersten militärischen Führers des Krieges*, ed. Eliza von Moltke (Stuttgart, 1922), p. 17.

[37] Moltke to Freytag-Loringhoven, 26 July 1915. BA-MA, W-10/51063.

[38] Ibid.

intention, despite the Chief of Staff's unambiguous statements, and despite the evidence contained in the prewar mobilization orders.[39] It is undeniable that the march through Belgium was an integral part of the German deployment plan, and that it was to be implemented regardless of the actions of the opponents.

The deployment plans of 1908–09 mentioned for the first time that troops should aim not to touch "the territory that immediately borders on Holland," and that their ability to do so would depend on Belgian and Dutch actions.[40] From that date onward, a march through The Netherlands was only planned in case Moltke's surprise attack on Liège failed.[41] The deployment plan for 1909–10 still included routes on Dutch territory "in case Liège cannot be taken early enough."[42]

There is much further evidence of Germany's oft-stated intention to violate Belgian neutrality. Bavarian Envoy Count Hugo von Lerchenfeld, to give but one example, reported on 5 August 1914 Moltke's reaction to an apparent offer of British neutrality if Germany refrained from an offensive in the West:

> An attack from German territory would have cost the German army 3 months and would have ensured Russia such a head start that we could no longer have reckoned on a success on both fronts. We had to go via Belgium with all our might on Paris for a quick reckoning with France. This was the only way to victory.[43]

In a study on the "deployment and operational intentions of the French in a future German-French war," the General Staff had concluded already

[39] Without providing any references for his statements, Zuber claims that the German Secret Service (!) decided to take Liège because they had concluded that the British were determined to support the French. "Der Mythos vom Schlieffenplan," in Ehlert et al., *Schlieffenplan*, p. 66. For a detailed critique of his position see Mombauer, "Of War Plans," in ibid., pp. 866–76.

[40] "Aufmarschanweisungen für die Jahre 1893/94 bis 1914/15," BA-MA, RH61/v. 96. Interestingly, in the maps that accompany these war plans (which were drawn after the war), Belgium is not even marked as a neutral country from 1908–09 onward.

[41] "Aufmarschanweisungen," BA-MA, RH61/v. 96, 1908–09 onward. See also Ehlert et al., *Schlieffenplan*, pp. 426ff.

[42] "Aufmarschanweisung," BA-MA, RH61/v.96, 1909/10. An interesting point of detail to emerge from these documents is that the German war plan had not dispensed with a violation of Dutch territory entirely, even for 1914. The deployment plan of 1913–14 stated: "For the removal of Liège is planned: in the night from 4th to 5th mobilization day a *coup de main* with previously dispatched 5 mixed infantry brigades.... In case of failure a 2nd *coup de main* with 3 inf. div... around the 10th mobilization day. This is to be combined with a *coup de main* on Huy. If Liège has not fallen by the 12th mobilization day, the 1st Army has to step on Dutch territory in its advance." BA-MA, RH61/v.96; also Ehlert et al., *Schlieffenplan*, pp. 467–77.

[43] Ernst Deuerlein, ed., *Briefwechsel Hertling-Lerchenfeld 1912–1917* (Boppard, 1973), p. 330.

in 1912 that the French would not "deploy along the Belgian border. Their right flank would then be threatened too much from the direction of Lorraine."[44] Such documents provide convincing evidence against the claim that German troops marched into Belgium to preempt a French offensive through Belgium.

As a result of Moltke's determination to safeguard Dutch neutrality, the infamous surprise attack on Liège became an integral part of the Western deployment plan.[45] In contrast, Schlieffen had never intended to take Liège; his plan had still relied on marching through The Netherlands, in which case Fortress Liège would not have been as crucial an objective. Only when Moltke considered it more prudent to avoid violating Dutch neutrality did capturing Liège become crucial, and only then did the bold stroke against the fortress become a military necessity.

A further difference between Schlieffen and Moltke was that the latter could no longer simply ignore the East, as Schlieffen was able to do in 1905 when Russia had been weakened following the Russo-Japanese War.[46] A few years after that war, Russia's military situation was much improved, and Moltke felt that he needed to deploy more troops there than Schlieffen had considered necessary in 1905.[47] As Bruce Menning argues in his chapter, better relations between Russia and France also played a part in this development, particularly as the two allies improved their plans for military cooperation in case of a war against Germany.[48] Moltke could no longer rely on Russia being weak, or even necessarily slow. In a 1912 memorandum (retirement was no obstacle to his war

[44] BA-MA, W-10/50267 (RH61/398), p. 124.

[45] This can clearly be seen in the "Aufmarschanweisungen" of the prewar years. Ehlert et al., *Schlieffenplan*, pp. 432–84.

[46] For a detailed argument to that effect see Robert T. Foley, "Der Schlieffenplan. Ein Aufmarschplan für den Krieg," in Ehlert et al., *Schlieffenplan*, pp. 101–16.

[47] See Schlieffen's memoranda of 1905 and 1912 in Ritter, *Schlieffenplan*, pp. 141ff. A copy of his memorandum of 28 December 1912 can be found in BA-MA, N121/35.

[48] Zuber also claims that the French, far from reacting to German aggression, only acted in order to honor their alliance obligation with Russia. Zuber, *Inventing*, p. 264. According to Zuber, it was not the Germans who had the offensive war plan, but the Franco-Russian alliance. On the French war plan, see Robert A. Doughty's chapter in this book, and "French Strategy in 1914: Joffre's Own," *Journal of Military History* 67 (April 2003): 427–54; also Stefan Schmidt, "Frankreichs Plan XVII. Zur Interdependenz von Außenpolitik und militärischer Planung in den letzten Jahren vor Ausbruch des Großen Krieges," in Ehlert et al., *Schlieffenplan*, pp. 221–56. On Russia's plan see Bruce Menning's chapter in this book; and Jan Kusber, "Die russischen Streitkräfte und der deutsche Aufmarsch beim Ausbruch des Ersten Weltkrieges," in Ehlert et al., *Schlieffenplan*, pp. 257–68.

planning!) Schlieffen had still confidently predicted that "Austria's fate will not be decided on the Bug, but on the Seine." In contrast, by 1914, Moltke feared that "under current conditions there can be no talk of Russia perhaps hesitating to advance into Prussia."[49]

This realization led to the scrapping of the Eastern Deployment Plan, which for a few years had been considered as an alternative to the Western Deployment Plan. In April 1913 it was decided that it would be abandoned.[50] From then on, Germany had only one war plan which it would have to implement regardless of the *casus belli.* The *Mobilmachungsterminkalender* (mobilization schedule) for 1913–14 and 1914–15 thus stated clearly: "*Only one deployment* is prepared in which the German main forces deploy on the *western front against France.*"[51]

The schedule called for only initial defensive action against Russia, a task that was to fall to Eighth Army. It was "to guard our Eastern provinces against a Russian invasion, [and] to support the offensive that Austria intends." For 1913–14, it was even noted that "the 8th Army should be deployed in the West if its deployment against Russia is not necessary."[52]

By the following year, that instruction was slightly modified. "If Russia remains neutral permanently the 8th Army will perhaps be used against France. Its dispatch [*Abtransport*] is prepared."[53] These mobilization instructions show clearly that the General Staff no longer counted on Russia's slow mobilization, which would have enabled Germany to ignore Russian troops almost completely in the first weeks of the war. The deployment of an entire army was deemed necessary. However, the possibility that these troops might not even be needed in the East had at least been considered, as the mobilization schedules clearly show.

[49] Moltke's marginal notes on Schlieffen's memorandum of 28 December 1912; Ritter, *Schlieffenplan*, p. 185; BA-MA, N121/35, marginal comments, p. 5.

[50] The deployment plans prior to 1909 contained a deployment West and East. From 1909–10, they contained Deployment II "Grosser Ostaufmarsch" against Russia. For the deployment of 1913–14 it was noted: "Germany's war preparations must primarily be directed against France.... We cannot reckon on a German war with England or Russia on their own given French bad feelings [*die französische Verstimmung*]." The "Grosser Ostaufmarsch" of the previous years now no longer existed. BA-MA, RH 61/v.96; Mombauer, *Moltke*, pp. 100–05.

[51] "Auszug aus dem Mobilmachungsterminkalender," for 1913–14 and 1914–15, in Ehlert et al., *Schlieffenplan*, pp. 468, 479. Original emphasis.

[52] "Aufmarschanweisungen für 1913/14," BA-MA, RH 61/v.96; Ehlert et al., *Schlieffenplan*, p. 468.

[53] Ibid, p. 479.

Another difference between the two plans resulted from the fact that Moltke needed to be more flexible in the West than Schlieffen had intended. Intelligence information showed that the French had abandoned their previous defensive stance and might favor an offensive in case of war. Moltke had already annotated Schlieffen's memorandum of 1905 with the comment that he "did not consider it a given that France would in all circumstances act defensively."[54] As a result, it might be necessary to deploy more troops in the South than Schlieffen, assuming a French defensive there, had intended in 1905.[55] Moltke noted:

> There is no point to continuing to march through Belgium with strong forces if the main French army advances in Lorraine; then only one thought can be decisive: to attack the French army with all available forces and to defeat it where it can be found. The march through Belgium is therefore not an aim in itself [*Selbstzweck*], but only a means to an end.[56]

In Moltke's view, defensive thinking was neither in keeping with the "inherent offensive spirit" of the French nation, nor "with the currently held teachings and views in the French Army," a conclusion attested to by Robert Doughty in this volume.[57]

As a result of such views, the great General Staff exercise of 1906 was already based on an assumed French offensive in the South. In such a scenario, as Moltke explained in his exercise critique, the German right wing would not be able to achieve a decisive victory, given that the majority of French troops would be deployed elsewhere.[58] The deployment plan of 1908–09 envisaged for the first time the deployment of one army corps to defend the Upper Alsace.[59] From this year onward, Seventh Army was to be deployed on the Rhine to defend Alsace, while Sixth Army was to be deployed in Lorraine; and from 1910–11 onward, it was no longer planned to deploy Seventh Army elsewhere.[60]

[54] Reichsarchiv, *Weltkrieg*, vol. 1, pp. 62–63. See also Ritter, *Schlieffenplan*, p. 145.

[55] Samuel R. Williamson, Jr., "Joffre Reshapes French Strategy," Paul M. Kennedy, ed., *The War Plans of the Great Powers, 1880–1914* (London 1979), p. 135; Doughty, "French Strategy in 1914," pp. 427–54.

[56] Cited in Ritter, *Schlieffenplan*, p. 56.

[57] Ibid., pp. 145–46; Reichsarchiv, *Weltkrieg*, vol. 1, pp. 62–63.

[58] "Aus der Schlussbesprechung der grossen Generalstabsreise 1906," BA-MA, W-10/50897; "Aeusserungen des Generals von Moltke über die Möglichkeit einer schnellen Feldzugsentscheidung im Westen," ibid., p. 138.

[59] "Aufmarschanweisungen" in BA-MA, RH61/V.96; Ehlert et al., *Schlieffenplan*, pp. 426–31; Reichsarchiv, *Weltkrieg 1914–1918*, vol. 1, p. 61.

[60] BA-MA, W-10/50730, p. 56. Copies of all the deployment plans can be found in Ehlert et al., *Schlieffenplan*.

The German war plan for 1913–14, the one which was implemented in August 1914, thus resulted in the following troop movements: seventy divisions to march immediately to the West and nine divisions to remain in the East. Of the remaining reserve of eight and one-third divisions (of which six and one-third were Ersatz divisions in the process of being formed), Moltke wanted to use five divisions in the East, raising the total there to fourteen. The remaining three and one-third were to remain in Schleswig-Holstein, there to guard the German North Sea coast, and even the shores of Denmark and The Netherlands, against possible British landing attempts. Moltke feared that the next war might even have to be fought on three fronts. In the West the total number of divisions was seventy-three and one-third. These were divided into seven armies.[61] This is how Germany went to war in August 1914.

The Great Naval Game

Neither Schlieffen nor Moltke included the Imperial German Navy in their war plans.[62] In the Navy Bills of 1898 and 1900 as well as Supplementary Bills (*Novellen*) of 1906, 1908, and 1912, Alfred von Tirpitz foresaw the creation by the mid-1920s of a battle fleet of sixty-one dreadnought battleships and battle-cruisers, to be deployed against Britain in the waters "between Helgoland [Island] and the Thames" River. For Tirpitz, Germany's future lay on the ocean by way of colonies and overseas trade. He saw Britain as "the enemy against which we most urgently require ... naval forces as a political power factor."[63] In what historian Paul Kennedy has called the "dagger at the throat strategy," the admiral argued that this mighty host of capital ships would force "perfidious Albion" to offer Germany "concessions" in the way of bases, colonies, and markets, rather than risk an all-out naval Armageddon

[61] First Army (six corps), General von Kluck; Second (six corps), General von Bülow; Third (four corps), General von Hausen; Fourth (five corps), Duke Albrecht of Württemberg; Fifth (five corps), German Crown Prince; Sixth (five corps), Crown Prince Rupprecht of Bavaria; and Seventh (three corps), General von Heeringen. In the East, Eighth Army (four corps), General von Prittwitz.

[62] The contours of German naval building have been outlined by Volker R. Berghahn, *Der Tirpitz-Plan. Genesis und Verfall einer innenpolitischen Krisenstrategie unter Wilhelm II.* (Düsseldorf, 1971); and Michael Epkenhans, *Die wilhelminische Flottenrüstung, 1908–1914. Weltmachtstreben, industrieller Fortschritt, soziale Integration* (Munich, 1991).

[63] Tirpitz's memorandum of 1 June 1897. BA-MA, F 2032/PG 6023 Reichs-Marine-Amt, Zentralabteilung.

in the south-central North Sea.[64] But if Britain should refuse such a generous course of action, Tirpitz was prepared to let the "Anglo-German naval race" be decided at sea, winner take all.

German naval strategy, for all its homage to Alfred Thayer Mahan's emphasis on the centrality of battle and on geographical position, in fact atrophied in the years before 1914. No clear strategic concept guided the "Tirpitz Plan."[65] The admiral simply insisted that the Reich build ships first, then craft a strategy for their deployment later. The British, he thought, on the basis of their national character and historical precedent, simply would "come out of port" to seek a "second Trafalgar" at the outbreak of any war. The notion that the British might be content to set up a distant blockade at the two outlets to the North Sea, the English Channel and the Norway-Scotland passage, against both Germany and the European neutrals was never seriously contemplated by Tirpitz. Thus, when Chief of the Admiralty Staff August von Heeringen already in 1912 predicted a "very sad role for our beautiful High Seas Fleet" should the British opt for a distant blockade, Tirpitz had no ready response.[66] Schlieffen and Moltke, for their part, saw no need to coordinate the nation's war plans with the Imperial Navy. Any British army that landed on the Continent, in their view, would simply be swept up by the great German "wheel" advancing through Belgium and France.

The Moltke Plan in Action: The Implementation of Germany's War Plan

The recent controversy around the German war plan of 1914 has at times assumed almost absurd characteristics, not least when the arguments advanced go counter to the events of 1914. Because we know what the German armies did in August 1914, we have a clear indication of what the war plan entailed. After all, it was implemented and, at least initially, seemed to be succeeding in leading the army to the desired swift

[64] Paul M. Kennedy, "Tirpitz, England and the Second Navy Law of 1900: A Strategical Critique," *Militärgeschichtliche Mitteilungen* 8 (1970): 38.

[65] For a recent analysis see Rolf Hobson, *Imperialism at Sea: Naval Strategic Thought, the Ideology of Sea Power, and the Tirpitz Plan, 1875–1914* (Boston, 2002). The basic documents are in Volker R. Berghahn and Wilhelm Deist, eds., *Rüstung im Zeichen der wilhelminischen Weltpolitik. Grundlegende Dokumente 1890–1914* (Düsseldorf, 1988).

[66] Cited in Holger H. Herwig, "The Failure of German Sea Power, 1914–1945: Mahan, Tirpitz, and Raeder reconsidered," *The International History Review* X (1988): 82.

victory. According to the historians of the Reichsarchiv, the mobilization in the West was "an extraordinarily remarkable achievement. The first great test of the reliability and thoroughness of the peacetime work of the German army could be considered passed."[67] Given that an entire "mobilization machine" was set in motion, in which approximately six percent of the population had to be transported, equipped, and kept supplied, the fact that this was achieved without any major difficulties was indeed an impressive accomplishment.[68]

As seen, it was part of the German war plan to advance quickly towards Belgium. With this in mind, neutral Luxembourg, whose railway lines were essential for the advance, was invaded on 2 August. The government there protested but offered no resistance.[69] The German advance into Belgium began on 4 August. The assault on Liège was the first major objective of the war. The ultimate success of Germany's strategic plan depended on its positive outcome.

However, the storming of one of the strongest fortresses in Europe was more difficult than anticipated. The General Staff had only expected about 6,000 garrison troops and 3,000 militia.[70] This proved to have been an underestimation. Manned with 35,000 garrison troops, Liège put up strong resistance, and the early mobilization of the Belgian army further impeded German hopes of quickly capturing the fortress town. As a result, it appeared initially as if the important bold stroke was failing. Erich Ludendorff, who as former head of the General Staff's Mobilization and Deployment Department knew the details of the plan, intervened. Under his leadership General Otto von Emmich's troops entered the city and captured the central citadel on 7 August, although not without suffering heavy losses, particularly among officers.[71] But the city's twelve reinforced concrete forts had to be overcome one by one. German troops were still fighting in Liège when the first British troops landed on the Continent on 12 August. The last of the forts was finally taken on 16 August, and only on the 17th, two days later than planned, could the right

[67] *Der Weltkrieg*, vol. 1, p. 137.

[68] Figures from Stefan Kaufmann, *Kommunikationstechnik und Kriegsführung 1815–1945* (Munich 1996), p. 145.

[69] *Der Weltkrieg*, vol. 1, pp. 105ff.; Josef Stürgkh, *Im Deutschen Grossen Hauptquartier* (Leipzig, 1921), p. 37.

[70] *Der Weltkrieg*, vol. 1, p. 108.

[71] Karl von Wenninger's report of 9 August 1914 attaches great importance to the Zeppelin bombing campaign in the success against Liège. He estimated German losses at that time between 1,000 and 3,000 men. BHStA-KA, MKr 1765, Nr. 2820.

wing of the German army begin its enveloping move through Belgium.[72] Liège had required "an expenditure of men and munitions beyond all expectations."[73]

Great relief was felt in Berlin when the initial rumors of a German defeat at Liège were refuted. The Kaiser was overjoyed. General Max Hoffmann recorded in his diary that taking Liège:

> was especially important. It had been long prepared for and the place had been thoroughly reconnoitered, so that it was depressing to learn that the first attack had failed. We were weeks behind in the entire campaign, and as everything depended on a speedy victory in the West, our joy over the success was doubly great.[74]

Liège's capture cleared the way for the advance of the right wing of the German army to meet its French and British opponents. Moltke now faced the problem of anticipating and responding to enemy action and predicting where the French would concentrate their troops. Would they send considerable numbers north to counter the German advance, or would they concentrate their forces in Lorraine to attempt an offensive in the South? From the middle of August onward, Moltke expected a French offensive between Metz and the Vosges where it was assumed that the French had amassed about one-fifth of their army.[75] The plan was to lure the French behind the Saar River and then launch a counterattack.

In three days of fighting, the German south wing (*Südflügel*), consisting of Sixth and Seventh Armies, managed to achieve victory in the Battle of Lorraine (20–23 August). Although they had not managed to envelop and destroy their opponent, the hasty retreat of the French was reported as a victory to Army Supreme Command (OHL).[76] When news

[72] Schlieffen had preferred avoiding Liège by marching through The Netherlands because he had anticipated that the fortress town would be difficult to take. Gerhard Ritter, *Staatskunst und Kriegshandwerk. Das Problem des 'Militarismus' in Deutschland* (4 vols., Munich, 1954–68), vol. II, p. 332. Georg von Waldersee maintained after the war that the Military Cabinet's decision not to deploy generals with knowledge of the area for the surprise attack was to blame for the initial failure to capture Liège. *Vossische Zeitung*, 18 and 25 September 1932. For details on the taking of Liège see Mombauer, *Moltke*, chapter 5.

[73] Sewell Tyng, *The Campaign of the Marne 1914* (New York and Toronto, 1935), p. 58.

[74] Max Hoffmann, *War Diaries and Other Papers* (2 vols., London, 1929); *Aufzeichnungen des Generalmajors Max Hoffmann* (Berlin, 1928), vol. I, p. 38.

[75] *Der Weltkrieg*, vol. I, p. 184.

[76] Tappen, "Vor zwanzig Jahren. Generaloberst v. Moltke," BA-MA, N56/5, pp. 262ff.; *Der Weltkrieg*, vol. I, pp. 263ff; p. 302.

of the victory arrived at headquarters on 22 August, the Kaiser's previously depressed mood lifted and the OHL was immediately led to exaggerated expectations. "The strategic hopes are very wide-ranging," Admiral Georg Alexander von Müller recorded.[77] General Hans von Plessen remembered after the war: "On 24th or 25th of August [Gerhard] Tappen literally said to me: 'The whole thing will be over and done with in six weeks'."[78]

At the same time, the armies of Germany's right wing had begun their advance on 18 August. Following the seizure of Liège, German troops moved rapidly into Belgium and France. En route for Paris, they reached Brussels on 20 August. Between 14 and 25 August, they engaged French and British forces in the so-called Battle of the Frontiers.[79] General Moriz von Lyncker, chief of the Military Cabinet, commenting on Karl von Bülow's victories on the right wing, noted that what had been planned "yet remained incredible" before the war had "begun to become reality." He recorded enthusiastically in his diary:

> There is no doubt in the General Staff that the operations of the 1st and 2nd Armies will lead to a good outcome and it is considered impossible that the English might still be able to escape us.[80]

For the German army, the war plan seemed to be delivering its promise of speedy and decisive victory, while for the French, the Battle of the Frontiers spelled the end of General Joseph Joffre's *Plan de renseignements* (the so-called Plan XVII).

In the South, Sixth and Seventh Armies managed to hold out and finally push back a strong French offensive. Because a military decision on the right wing seemed imminent, Moltke decided to engage the French further in Lorraine, rather than to move troops north to strengthen the right-wing advance. He aimed to drive the French armies apart. The right wing could then use its superiority to achieve an overall victory.[81]

77 Walter Görlitz, ed., *Regierte der Kaiser? Kriegstagebücher, Aufzeichnungen und Briefe des Chefs des Marinekabinetts Admiral Georg Alexander von Müller 1914–1918* (Göttingen, 1959), p. 50.

78 "Mündliche Mitteilungen des Generalobersten von Plessen," 10 April 1923, BA-MA, W-10/50897, p.148. For further optimistic statements from Plessen's diary, see Holger Afflerbach, ed., *Kaiser Wilhelm II. als Oberster Kriegsherr im Ersten Weltkrieg. Quellen aus der militärischen Umgebung des Kaisers 1914–1918* (Munich, 2005), pp. 643–56.

79 Herwig, *First World War*, p. 97.

80 Moriz von Lyncker, diary entries, 25 and 27 August 1914 in Afflerbach, ed., *Kaiser Wilhelm II.*, pp. 142, 144.

81 After the war, Rupprecht and Krafft von Dellmensingen stressed their own surprise when they received the order to engage the French further. While Krafft's postwar notes

Moltke's decision to pursue the enemy with the armies of the *Südflügel* amounted in effect to a strategy of "double envelopment." This has been interpreted as an adulteration of the Schlieffen Plan and as a serious misjudgment, both by contemporary critics and some historians. Historian Holger Herwig, for example, considers "Moltke's spur-of-the-moment decision to send the sixteen divisions of the German sixth and seventh armies not north via Metz to join Bülow and Kluck but east towards Épinal, ... one of his gravest errors during the Marne campaign."[82] Colonel Tappen argued after the war that transport problems had been the main reason for the decision not to move troops north to strengthen the right wing. With the Belgian railways largely destroyed, it would simply have taken too long for troops from the South to get there to have much of an impact on the ongoing fighting.[83]

The naval side of the war had also not gone well. The capital ships of the Royal Navy, reconstituted as the Grand Fleet, did not "come out" as predicted by Tirpitz, and instead instituted a distant blockade from their lairs at Rosyth and (later) Scapa Flow. Shipping to neutrals such as The Netherlands never provided the "windpipe" that Moltke had envisaged as the tough enemy blockade interdicted all neutral transport under search-and-seizure laws. Admiral von Heeringen's "beautiful High Seas Fleet" was reduced to playing the part of a classic "fleet-in-being," one too small to challenge Britannia on the open sea and yet one too large to risk in an all-out encounter. Interdiction of the transport of the British Expeditionary Force to France was not attempted.

The Failure of the Moltke Plan: The Outcomes of German War Planning

By 22 August, the invasion of Belgium was nearly completed. The fighting of 20–23 August had decided the Battle of the Frontiers in Germany's favor. The armies of the Entente were retreating. The German war plan seemed to be delivering its promised speedy victory in the West, to the

were perhaps marred by his intention to justify his actions, Rupprecht's diary confirms that they did not seek to continue pursuit of the French at that time. Crown Prince Rupprecht, *Mein Kriegstagebuch* ed. Eugen Frauenholz (Berlin, 1929), vol. I, p. 38.

[82] Herwig, *First World War*, p. 99.

[83] Tappen, "Kriegserinnerungen," BA-MA, W-10/50661p. 27; Gerhard Tappen, *Bis zur Marne 1914. Beiträge zur Beurteilung der Kriegführung bis zum Abschluss der Marne-Schlacht* (Oldenburg, 1920), p. 14.

delight of the German military leaders who confidently anticipated the entry of German troops into Paris in coming weeks.[84]

By the beginning of September, the German right-wing armies had almost reached the French capital. General von Lyncker wrote to his wife on 20 August: "[O]ur operations are progressing well. . . . We are already nearer to Paris with parts [of our army] than many French people."[85] Not surprisingly, almost believing themselves to have victory in their grasp, Germany's military and political leaders expressed far-reaching war aims. On 7 September, Moltke voiced his desire for a peace "which for the foreseeable future could not be disturbed by any enemy" – his privately uttered doubts of the prewar months that Germany was not superior to its enemies seemingly forgotten.[86] Two days later, in the infamous "September-Program," Chancellor Theobald von Bethmann Hollweg formulated his vision of Europe following a German victory, which – in the light of news of victory upon victory – he clearly thought to be imminent.[87]

And yet, a decisive victory against the French armies was not yet achieved. In private, the chief of staff still warned against over-optimism: "We don't want to delude ourselves. We have had successes, but we have not yet won." Moltke, not known for his optimism, declared: "The hardest task is still ahead of us!"[88] He was right. Within days, the German armies on the Western Front would be retreating from the enemy for the first time. Moltke had the task of ordering that retreat and thus abandoning the war plan which he and his predecessor had refined for years.

In the East, Germany's troops had suffered an initial setback in the Battle of Gumbinnen on 20 August. Given the encouraging news from the armies of the right wing in the West, and faced with the possibility of a Russian invasion of East Prussia and the threat this posed to Berlin, Moltke decided to move two corps from the Western to the Eastern front

[84] For details of the Battle of the Marne see Tyng, *Campaign of the Marne*; Mombauer, "Marne," *passim*.

[85] Moriz von Lyncker, letter of 20 August 1914; Afflerbach, *Kaiser Wilhelm II.*, p. 138.

[86] Moltke's citation in Fritz Fischer, *Griff nach der Weltmacht. Die Kriegszielpolitik des kaiserlichen Deutschland, 1914/18* (Düsseldorf, 1984), p. 90. On Moltke's previous pessimism, see Mombauer, *Moltke*, p. 211.

[87] Details on the "September-Program" in Fischer, *Griff nach der Weltmacht*, p. 90. But see also Wayne C. Thompson, "The September Program: Reflections on the Evidence," *Central European History* 9 (1978): 342, who claims that Riezler and Bethmann Hollweg did not expect a victory to be imminent.

[88] Moltke's conversation with Karl Helfferich on 4 September 1914. Karl Helfferich, *Der Weltkrieg* (2 vols., Berlin 1919), vol. 2, p. 18.

(the Guard Reserve Corps and XI Army Corps) to help the beleaguered Eighth Army. However, before the reinforcements arrived, Generals Paul von Hindenburg and Erich Ludendorff managed to avert a threatened German defeat and instead produced a major victory at the Battle of Tannenberg (27–30 August). Moltke's critics would later maintain that those two corps could have been useful in preventing the German retreat at the Battle of the Marne, but such is the benefit of hindsight.

Moltke's decision was made in view of the desperate situation of Eighth Army. In his diary, Hans von Plessen, commandant of the Kaiser's headquarters, recorded the bleak mood in the Supreme Army Command:

> But East Prussia! Gumbinnen and Allenstein occupied by the enemy! The Russians burn and pillage everything! – We must make haste to finish up in the West as quickly as possible in order to come to the rescue of the East.[89]

German military doctrine, as historian Dennis Showalter points out, had emphasized for the previous twenty years that troops needed to be sent to the East following initial, decisive victories in the West.[90] On 25 and 26 August this point seemed to have been reached, and the decision was made to move troops to help the struggling forces in the East, this based on years of planning that had predicted that German troops would be urgently needed in the East following the initial successes against the enemies in the West. Yet, none of the war planners had developed contingency plans in case their plan failed, in case victory in the West was not quick or even eluded them altogether. When the events of the Battle of the Marne put an end to the German advance in the West, there was no "Plan B."

In France, the armies of the German right wing were facing superior French numbers as well as the British Expeditionary Force on a front almost 300 kilometers wide. Following the rapid advance of the German right-wing armies, General Alexander von Kluck's First Army – contrary to orders from the OHL – out-marched Bülow's Second Army, crossed the River Marne and found itself within reach of Paris. This move was not only a clear case of disobeying superior orders, but also an *ad hoc* departure from the Moltke-Schlieffen Plan. As a result of Kluck's aggressive action, German First and Second Armies were on the point of colliding at the Marne. Above all, the balance of power on the extreme right wing

[89] Plessen diary, 24 August 1914, BA-MA, W-10/50676. Now published in Afflerbach, *Kaiser Wilhelm II.*, p. 648, who points out that reports of Russian atrocities were based on exaggerated accounts.

[90] Dennis E. Showalter, *Tannenberg: Clash of Empires* (Hamden, CT, 1991), p. 294.

had changed to Germany's disadvantage, with twenty German divisions facing thirty French and British. The French had a significant advantage by being able to use their railways, allowing them a more rapid deployment of troops and, crucially, of supplies to the front, while the Germans faced severe transport and supply problems.

Additionally, the German armies, like those of the other major powers, suffered from problems of communication in the dawning "radio age." Command and control remained with telegraph, telephone and dispatch riders. Nor did cryptographic systems exist that allowed wireless to be used effectively and securely. The OHL possessed but one radio set with which to communicate with army headquarters, and the latter, in turn, had not provided their corps commanders with wireless. The upshot was not only that Moltke was without contact to First and Second armies for forty-eight hours during the critical phase of the Battle of the Marne, but also, as historian John Ferris has argued, that the French and British intelligence branches were able between September and November 1914 to intercept about fifty German radio messages transmitted in plain language.[91] The intercepts allowed Entente forces insight into the German command system's collapse and the developing gap between First and Second armies at the Marne.

Finally, enormous marching distances had led to a serious reduction of the German army's fighting ability. First Army had marched 500 kilometers in thirty days, and Third Army 300 kilometers in two weeks.[92] The Austrian military observer on the Western Front reported after the event that the real reason for the German retreat was only in part the fact that the enemies had almost succeeded in breaking through the German lines, but rather lay in the "exhaustion of the troops" and the reduced fighting power "due to very high losses." He also noted that the French leadership was able "through very skillful use of the railways to achieve continual large troop movements behind their own front from the right to the left wing" and thus to put up "very considerable and often in places very superior forces" against the German right wing.[93]

On 5 September, French Commander-in-Chief Joffre ordered a counterattack with regrouped forces against the flank of First Army. As a result, Kluck was forced to retreat, now following an earlier order from

[91] John Ferris, ed., *The British Army and Signals Intelligence during the First World War* (Phoenix Mill, UK, 1992), pp. 3–5.

[92] Max von Hausen, *Erinnerungen an den Marnefeldzug 1914* (Leipzig, 1920), pp. 178–80.

[93] Stürgkh's report of 30 September, Kriegsarchiv (KA) Vienna, AOK Op. Abteilung, Deutsche Westfront, 555a: Berichte aus dem deutschen Hauptquartier 1914/1918.

Moltke which he had so far ignored. In the process of trying to repel the French attack, a large portion of First Army, two corps, was moved 100 kilometers to try and outflank French General Michel Maunoury's Sixth Army. As a result, a gap of almost 40 kilometers developed between First and Second armies, into which the enemy threatened to advance. Bülow and Kluck lost touch with each other and with the OHL. Partly due to inadequate communications, they no longer coordinated their actions.

With First Army in danger of being outflanked by the enemy and a threatening advance of French and British troops in the opening between First and Second armies, the gap, which had widened still further due to Bülow's decision to withdraw his exposed right wing, had to be closed. Consequently, First and Second armies were ordered to retreat behind the Marne on 9 September, even though they had apparently so far been victorious.[94] For Lyncker, recording his impressions at headquarters, 9 September was "the most critical day of the campaign so far."[95]

The other armies of the right wing had to follow two days later, and any hope of defeating the French or reaching Paris quickly had been dashed. This, in a nutshell, was the Battle of the Marne. It paralyzed Moltke's advance in the West and thus put an end to his war plan. It was the first major German retreat as well as a defeat, and its impact on morale at headquarters was understandably grave, although contemporaries did not yet view it with the seriousness they would after 1918.

On 11 September, Moltke, together with his adjutants Wilhelm von Dommes and Gerhard Tappen, traveled to the army headquarters of the right wing when it appeared as if the enemy might break through between Second and Third armies. Confronted with the situation as it presented itself, Moltke had to give the order to withdraw Third, Fourth, and Fifth armies in order to reestablish a united front. That night, he returned to OHL headquarters at Luxembourg a broken man. General von Lyncker recorded in his diary on 13 September: "Moltke is completely crushed by the events; his nerves are not up to this situation."[96]

[94] See Bericht der OHL zum Ende der Marne-Schlacht, 10 September 1914, in W. Bihl, ed., *Deutsche Quellen zur Geschichte des Ersten Weltkrieges* (Darmstadt, 1991), p. 63.

[95] Lyncker diary entry, 9 September 1914; Afflerbach, *Kaiser Wilhelm II.*, p. 162.

[96] Lyncker diary, BA-MA, W-10/50676, pp. 29ff; Afflerbach, *Kaiser Wilhelm II.*, p. 165. See also Crown Prince Rupprecht's account of 8 September: "Gen. v. Moltke appears to me like a sick, broken man. His tall figure was hunched, he looked incredibly worn out." BA-MA, W-10/50659, p. 207; Rupprecht, *Kriegstagebuch*, vol. I, p. 103.

In his diary entry of 10 September, Bavarian Military Plenipotentiary Karl von Wenninger recorded the depressed mood at headquarters: "It is as quiet as a mortuary in the school building in Luxembourg – one tip-toes around, the General Staff officers rush past me with their eyes down-cast – best not to address them, not to ask."[97] There could be no starker contrast to the optimism and the "beaming faces" that he had recorded during the first days of war when Germany's military leaders felt they had "jumped the ditch."[98] None of the military planners in the OHL could have had any doubt that a retreat on the right wing not only went completely against their prewar planning, but more to the point, that there was no other plan which could now be implemented. In short, Germany's military planners had run out of ideas and from now on would have to react to the will of the enemy rather than trying to impose their own will on them.

Unlike the preceding victories, the Battle of the Marne was kept a secret at home, and even the Austro-Hungarian ally was not informed about it. What would come to be regarded as the dramatic events that seemed to mark the beginning of the end for Germany's and Austria-Hungary's war effort seemed much less enormous at the time. They could still be described quite soberly by General Josef Stürgkh, the Austro-Hungarian military observer at the Western Front:

> Due to advance of the 1st Army against Paris a hole developed between it and the 2nd Army, into which the opponent, particular the English, entered. Necessitated by this retreat of 1st Army and right wing of 2nd Army to the Marne, 3rd and 4th Armies remained in position, 5th Army achieved successful attack southwest of Verdun on 10 September. Extraordinary marches and the almost daily battles until now have caused very grave losses, for whose replacement a pause now seems necessary. Securing of communications and occupation of Belgium tie up significant forces. Currently impossible to extricate troops from the western theater of war.[99]

Moltke wrote only a single paragraph describing his order to withdraw Third, Fourth, and Fifth armies on 11 September: "It was a difficult decision that I had to take without being able to get His Majesty's agreement first. It was the hardest decision of my life, which made my heart bleed.

[97] Bernd-Felix Schulte, "Neue Dokumente zum Kriegsausbruch und Kriegsverlauf 1914," *Militärgeschichtliche Mitteilungen* 25 (1/1979): 172.

[98] Wenninger's report, 31 July 1914, in ibid., p. 140.

[99] KA Vienna, AOK Op. Abteilung, R.Gruppe: September 1914: 479. It should be pointed out, however, that the plenipotentiary was not being kept well informed and failed to gather much information on the events on the Western Front.

But I anticipated a catastrophe if I had not withdrawn the army."[100] In other words, while the outcome of the battle was disappointing, not to order the retreat would have had far graver consequences.

While Moltke clearly connected his dismissal on 14 September with his order to retreat, he did not give the event any more importance than that: it was a defeat on the right wing, which was lamentable, but more than anything else it was his own personal defeat, an event from which he would never properly recover. In the immediate aftermath of the battle, the retreat from the Marne did not seem to have meant the loss of the war, although it was considered a severe setback. It would take Germany's total defeat in 1918 to elevate both the events of September 1914 and the Schlieffen Plan to truly mythical status. With hindsight, however, it is clear that the Battle of the Marne spelt the end of the Schlieffen-Moltke Plan. The goal of a short war against one of the two major enemies, France, had not been realized and was now impossible to achieve.

Certainly, as historian Hew Strachan has argued, the Battle of the Marne had serious immediate political consequences. "France and the French army were saved: without that the Entente would have had no base for continuing operations in Western Europe." And there were serious longer-term strategic consequences as well: "Germany had failed to secure the quick victory on which its plan rested. From now on it was committed to a war on two fronts." While the Battle of the Marne was "tactically indecisive," Strachan argues that "strategically and operationally the Marne was a truly decisive battle in the Napoleonic sense." He concludes:

> The French had "fixed" the Germans in the east and manoeuvred to strike against them in the west; the Germans' initial victories had been valueless because they had neither fixed nor destroyed their opponents, but left them free to manoeuvre and to fight again.[101]

The events of September 1914, the first major German defeat, destroyed all hopes of the Schlieffen-Moltke Plan fulfilling its promise of swift victory in the West, although arguably a short war, even just in the West, had already become unlikely, given other military and diplomatic developments, such as the "Pact of London" of 5 September, wherein the governments of France, Great Britain, and Russia pledged that none of

[100] Moltke, "Die 'Schuld' am Kriege," in Moltke, *Erinnerungen*, p. 24.

[101] Hew Strachan, *The First World War*, vol. I, *To Arms* (Oxford, 2001), p. 261.

them would conclude a separate peace.[102] The military planners' worst nightmare, a long drawn-out war of attrition, a stalemate of trench warfare in which opponents of similar strength faced each other indefinitely, had become reality. German war planners had no contingency plans or alternatives to their one deployment plan which relied on a victory against France so as to be able to move troops to the East to deal with Russia, potentially an even more dangerous enemy.

Thus, the difference between plan and reality could not have been greater in 1914. The German offensive, intended to be swift and decisive, got bogged down outside Paris. Instead of a short, victorious war, Germany now had to fight a protracted war on two fronts which ended not in glorious victory, but in bitter defeat. This was not what any of Germany's military planners had in mind when the war that many had eagerly awaited had finally become reality. Although Schlieffen and Moltke (as well as the Elder Moltke before them) had at times voiced concern over a long future war, none appear to have imagined this outcome.

Seen many years after the events of 1914, perhaps the greatest paradox about the Schlieffen Plan is this: it was a war plan that was flawed in many ways and that failed to deliver victory; and yet, in popular (and military) memory in the interwar years, it became the stuff of strategic genius. It resonated well beyond 1914, for in World War II Germany attempted again (this time with a "three-dimensional Cannae") to achieve a quick victory in the West by initially using much the same "recipe for victory."[103]

Finally, in the light of the foregoing discussion of perhaps the most widely known and discussed war plan of the modern era, it is both irony and nemesis that the last chapter in Imperial Germany's war was to be written by the navy. Irony insofar as the High Seas Fleet had not made a decisive contribution to the war effort 1914–1918. Nemesis insofar as the fleet had poisoned relations with Britain and had cost roughly 800 million Goldmark to build, money that otherwise might well have been spent on expanding and modernizing the army and thereby given it a better chance to execute the Moltke Plan at the Marne.

[102] See L.L. Farrarr, Jr., "The Strategy of the Central Powers," in Hew Strachan, ed., *The Oxford Illustrated History of the First World War* (Oxford, 1999), p. 29.

[103] Hans Ehlert, Michael Epkenhans and Gerhard P. Groß, "Einleitung," in Ehlert et al., *Schlieffenplan*, p. 12.

For much of the war, the High Seas Fleet remained a "fleet-in-being," content to undertake several irritating but otherwise insignificant "tip-and-run" raids on the British east coast. After the strategically indecisive battle with the Grand Fleet off Jutland on 31 May–1 June 1916, German naval leaders on 1 February 1917 turned to unrestricted submarine warfare to bring Britain to its "knees" within six months. While scoring spectacular initial results – reaching the prescribed monthly total of 600,000 enemy tons sunk per month in March and topping it by 241,000 tons in April – the campaign ended in failure after Britain had instituted convoying of merchant ships and developed a highly efficient anti-submarine campaign. It had also brought the world's greatest neutral, the United States, into the war against Germany. That event alone perhaps was a fitting epitaph in general for German war planning 1914–1918.

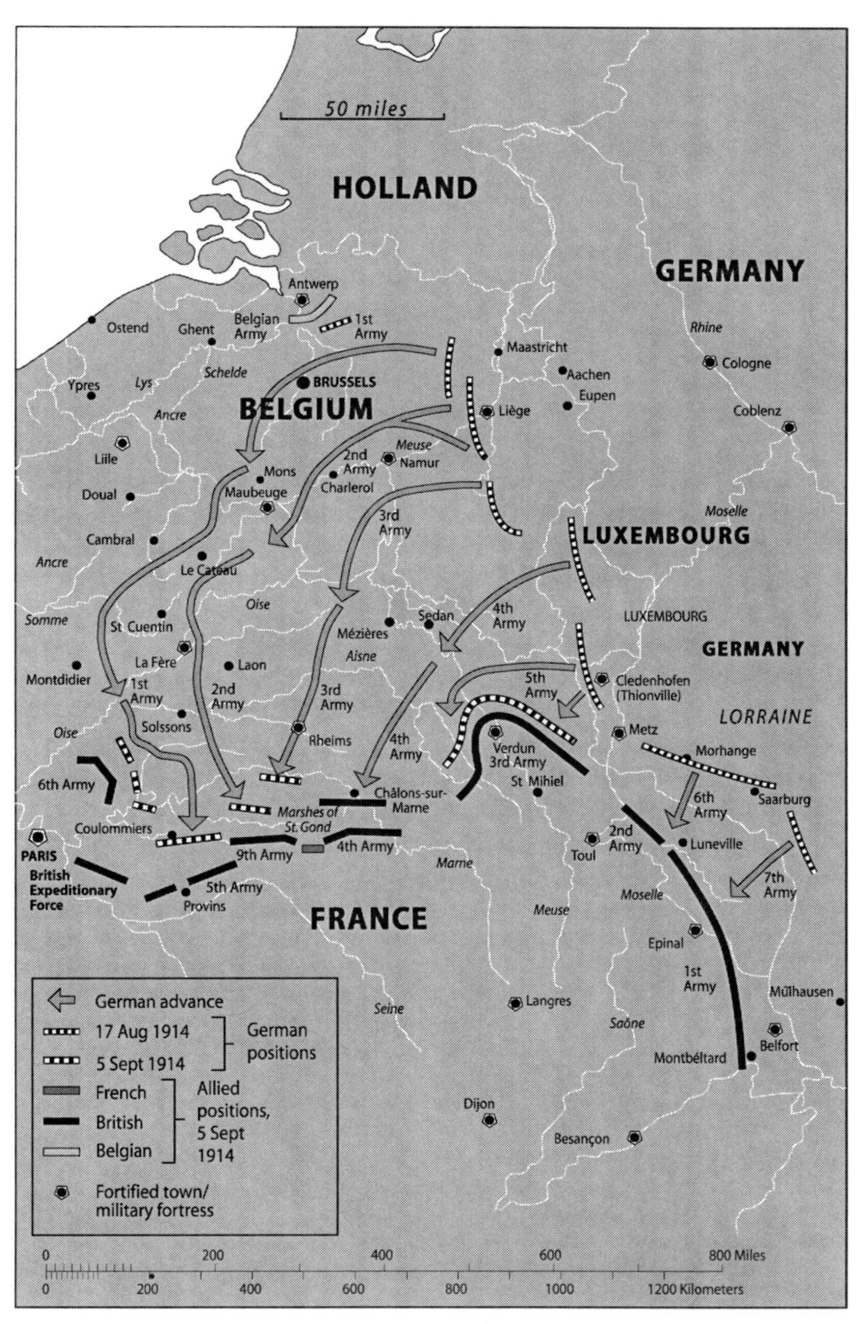

MAP 1. The German Advance, 1914.

4

War Planning and Initial Operations in the Russian Context

Bruce W. Menning

Between 1873 and 1914, Russian preparation for a European war evolved through two stages. The first was very long and the second very brief, and there was a jagged transition between the two. Some knowledgeable observers termed these stages "systems," because in varying degrees each period gave rise to an overarching scheme with a governing strategy that flowed from geopolitical circumstance, threat assessment, alliance commitments, course of action development, detailed planning, and resource allocation. For various reasons, neither period's system fully met requirements for engagement in a Great Power war. However, only the second, sometimes called the "Sukhomlinov System" after Russia's last pre-1914 war minister, would be put to the test.[1] When war did come, this system's inadequacies and internal inconsistencies meant that Russia would enter hostilities as it had entered nearly all its past wars, like a chess player with poor opening moves. The result during the initial period of conflict

[1] Two examples of the penchant to discern system are N. N. Golovin, *Iz istorii kampanii 1914 goda na russkom fronte. Galitsiiskaia bitva* [hereinafter Golovin, *Galitsiiskaia bitva*] (Paris, 1930), 25, and A. M. Zaionchkovskii, *Podgotovka Rossii k imperialisticheskoi voine* (Moscow, 1926), 141–54. The argument usually turns on issues of engineering infrastructure, especially fortress modernization, followed by reference to elements of strategy, force structure, deployments, and defensive-offensive courses of action.

With respect to conventions, all dates in this chapter are rendered according to the Julian calendar, which lagged behind the Gregorian by 13 days at the beginning of the twentieth century. Exceptions are noted either as N. S. (New Style) or with parenthetical reference to the corresponding Gregorian-style dates. All transliterations from Russian are rendered according to the modified U. S. Library of Congress system, except for proper and place names that have come into common English usage through other systems, e.g., Nicholas, Alexander, St. Petersburg, Moscow.

was catastrophic failure against Germany and only partial success against Austria-Hungary.

Setting the Terms

These outcomes testified to anomalies and dysfunctions that lay below the surface of the apparent system. In Russia, prevailing notions of statecraft and monarchical prerogative often inhibited the effective linking of ends, ways, and means according to a mature understanding of higher, or grand, strategy. In institutional terms, modified autocratic politics defined politico-military jurisdictions and the limits of higher-level coordinating authority in ways that precluded optimal collaboration during strategy development, war planning, and resource allocation. In direct planning terms, neither the Main Staff nor its post-1905 successor, the Main Directorate of the General Staff (Russian acronym *GUGSh*), consistently afforded the organizational framework required to cultivate competence and to extend collaboration. In conceptual terms, war planning as a process remained ill-defined and incompletely developed, with the result that officials and officers possessed only a rudimentary sense of what constituted a "war plan." They understood the war plan as an elaboration of larger strategic "considerations" (*soobrazheniia*) that constituted an operational concept, combined with a small family of schedules and lesser plans that served the concept.[2] The most important member of this family was the troop mobilization schedule, and it was so central to Russian strategic design that a given schedule was often misunderstood as synonymous with a war plan. In 1914, Russia went to war in accordance with a modified version of Mobilization Schedule No. 19A.

The first mobilization schedule dated to the 1870s, when Russia began the pilgrimage from a large standing ground force to a cadre- and reserve-based army. It was also during this decade that planners initially contemplated the possibility of conflict against the combined forces of a united Germany and a revitalized Austria-Hungary. Memories of the Crimean War of 1853–56 were still fresh, including two of its more poignant lessons: the vulnerability of a vast imperial periphery and the difficulties of waging war against a coalition.[3] In 1879, when Germany

[2] Cf., A. A. Neznamov, "Plan voiny," *Voennaia entsiklopediia*, 18 vols. incomplete (St. Petersburg/Petrograd, 1911–15), XVIII, 446–49, and Iu. N. Danilov, "Nashe strategicheskoe razvertyvanie v 1914 godu i idei, polozhennye v osnovu ego," *Voennyi sbornik* [Belgrade], no. 4 (1923), 73–4.

[3] William C. Fuller, Jr., *Strategy and Power in Russia, 1600–1914* (New York, 1992), 267.

and Austria-Hungary forged the Dual Alliance, Russia once again confronted the nightmarish prospect of a coalition-based threat to its imperial periphery.

Although the locale for potential confrontation had now shifted from the south to the west, the military-strategic situation only repeated a familiar dilemma under an altered and more modern guise. The fundamental problem was that Russia's primary potential foes were situated opposite the western military frontier, contiguous with tsarist borderlands in which large segments of the population were of uncertain loyalty to St. Petersburg. Meanwhile, Russia's primary strength, its pool of trained military manpower, was distributed unevenly across the imperial heartland, European Russia. For various reasons, the strategic railroad network that might have matched resources with threat was inadequate. This deficiency unavoidably subjected Russian strategists and planners to tyrannies of time and space. These the Russian military measured in days, by 1914-style calculation, roughly seven to fourteen.[4] Depending upon which formations were counted, when, and where, the varying number of these days marked the differential advantage held by Germany and Austria-Hungary over Russia in speed of troop mobilization, transit, and concentration for strategic deployment along the state frontier.

There were several ways to address the disadvantageous time-space-mass calculus inherent in Russia's strategic dilemma. One answer was to share the threat, to distribute Russia's wealth in enemies, and to give at least one of them pause for thought. In 1892, autocratic Russia did the heretofore unthinkable by concluding a military convention with republican France. In 1894, the convention became a defensive alliance. The arrangement provided for immediate and simultaneous French and Russian troop mobilization in response to mobilization by any or all signatories to the Triple Alliance (Italy had joined the Dual Powers in 1882, making them collectively the "Triplice"). Moreover, if Germany attacked France with at least two-thirds of its forces, Russia was to attack Germany "with utmost dispatch" with 700,000–800,000 "fully constituted" troops. Similarly, if Germany attacked Russia with at least two-thirds of its forces, France was immediately to attack Germany.[5] However, there were several major caveats to the arrangement that persisted until 1912: Russia subscribed to no timetable for offensive operations and

[4] N. G. Vasil'ev, *Transport Rossii v voine 1914–1918 gg.* (Moscow, 1939), 26.

[5] V. A. Zolotarev, Iu. V. Kudrina, et al. (eds.), *Mirovye voiny XX veka*, 4 vols. (Moscow, 2002), II, 13.

reserved the right in any case to retain a free hand in sending the bulk of its forces against either Germany or Austria-Hungary.[6] St. Petersburg contended with Vienna for Balkan advantage, and the Romanovs had scores to settle with the Habsburgs that dated to the Crimean War. Whatever the interpretation of alliance obligations, Russian planners always remained uncertain whether Germany might turn east or west at the onset of a European war. For Russia, the difference was significant. It was plus or minus roughly thirteen German corps, or perhaps as many as six field armies.[7]

These and other uncertainties underscored the importance of a general-staff style analysis of Russia's adversaries and the potential theater of war. The empire's western land frontier stretched an eye-popping 2,000 kilometers from Scandinavia in the North to the border with Romania and the Black Sea littoral in the South. Approximately half this distance delineated shared state boundaries with Germany and Austria-Hungary. Opposite these two potential adversaries, the immense western theater of war and its frontier trace necessitated a north-to-south consideration of three possible sub-theaters: northwest, "forward," and southwest.

Of the three, the forward theater, comprising the majority of Russian Poland, elicited the gravest concern. It constituted a prominent salient, full of peril and promise. Like the two sub-theaters on its flanks, the forward theater lacked natural barriers, save the Vistula River and its tributaries. Unlike its flanking neighbors, the forward theater's distant rear area was a dense swampy morass, the Pripet Marshes. Transited east-west by a lone railroad with reduced carrying capacity, these marshes bisected the entire potential western theater of war. Even after the railroad buildup at the end of the nineteenth century, only four double-track lines fed into the forward theater, and they were poorly connected by lateral feeder lines in the region west of the Pripet, between Brest-Litovsk and Warsaw. At the outset of possible hostilities or shortly thereafter, all these circumstances meant that substantial Russian troop deployments within the forward theater might be subject to potential entrapment. However, the same circumstances also meant that the same deployments – properly defended and reinforced – would be well poised for attack into either Austrian

[6] A. M. Zaionchkovskii, *Podgotovka Rossii k mirovoi voine v mezhdunarodnom otnoshenii* (Leningrad, 1926), 169–72.

[7] Rossiiskii gosudarstvennyi voenno-istoricheskii arkhiv [hereinafter RGVIA], *fond* [collection] 400 (*Glavnyi shtab*), *opis'* [inventory] 4, *delo* [item] 433, *list* [folio] 59 *obratnaia* [obverse side].

MAP 2. Russia: Potential Western Theater of War.

Galicia or German East Prussia – or perhaps even across Silesia into the heart of Germany.[8]

Demography also figured prominently in the military calculus for the forward theater. Planners understood that at best much of the population of Russian Poland simply tolerated its imperial masters as occupiers. If this territory were ever ceded to an enemy, then the Russians would have to fight twice, once to eject the invaders and again to subdue the Polish populace. It was for this reason that substantial Russian contingents deployed within the Polish salient during peacetime served a dual purpose, as defenders and occupiers.[9] This situation underscored the understanding that Russia's greatest strategic asset was reliable military manpower from the interior.

The key element of organizational infrastructure that linked requirements in the West with resources in the interior was the Russian system of military districts. At the outbreak of war in 1914, they numbered 12, with eight in European Russia and four in Asiatic Russia. A product of the military reforms during War Minister D. A. Miliutin's tenure (1861–1881), each military district headquarters resembled a miniature war ministry with major peacetime and wartime responsibilities. With the advent of a universal military service obligation in 1874, it was the function of the military districts to recruit, train, and assign troops to active or reserve formations, some of which resided in the districts themselves.[10] Between 1890 and 1914, the district chiefs and their staffs were slated to assume important wartime command and staff functions. Those three military districts (Kiev, Warsaw, and Vilnius) that shared borders with Germany and Austria-Hungary would provide the commanders and staffs for the two potential wartime fronts (Northwest and Southwest) against Russia's primary adversaries.

A Russian front was the equivalent of a western-style army group, but with significant authority over immediate rear areas. For planning purposes, the chief of the Kiev military district usually migrated with his staff to become commander of Southwest Front against Austria-Hungary, while the commander of Northwest Front against Germany came from either the Warsaw or Vilnius military districts. Chiefs of prominent

8 Ibid., d. 445, ll. 75–7; see also, Douglas Wilson Johnson, *Topography and Strategy in the War* (New York, 1917), 50–61.

9 Jeffrey Aaron Mankoff, "Russia and the Polish Question, 1907–1917: Nationality and Diplomacy," Ph. D. diss., Yale University, 2006, 116–17, 129–30.

10 N. N. Golovin, *Iz istorii kampanii 1914 goda na russkom fronte. Plan voiny* [hereinafter Golovin, *Plan voiny*] (Paris, 1936), 4–5.

interior military districts, for example, Moscow and Kazan, usually acceded to the wartime commands of armies drawn from their districts. On the extreme flanks of the western theater of war, the chiefs of the St. Petersburg and Odessa districts usually became the wartime commanders of separate armies assigned to flank security.[11]

In accordance with the Russian system of troop mobilization, it was the function of the mobilization schedule to match troops from various military districts with unit deployments near and far. The process was complex, beginning with assembly and proceeding through transit to concentration and beyond to strategic deployment. Compilation of schedules required more than two years, and the advent of each new version corresponded roughly with the advent of each new five-year planning and budget cycle.[12] A single schedule comprised several hefty tomes, and its components were sufficiently complex and interwoven so that improvisation at any juncture during implementation became an invitation to confusion and disaster. Documentation of the many constituent parts was so crucial that there could be no plan for war without a troop mobilization schedule. At the outset of hostilities in 1914, full Russian troop mobilization expanded the active field force from 1.3 to 4.7 million men under arms and drafted 1.1 million horses.[13]

Engineering infrastructure, especially railroad networks and fortress complexes, formed the iron and masonry skeleton on which troop mobilization schedules hung human muscle and horse flesh. The inherent vulnerability of the Polish salient to envelopment underscored the necessity for a system of fortresses and fortifications to guard key terrain features, population nodes, and potential areas for troop concentrations. Therefore, the Polish salient and its immediate rear area until 1910 counted a network of seven fortresses, four of which (Żegrze, Novogeorgievsk, Warsaw, and Ivangorod) constituted a forward shield astride the middle Vistula and lower Bug. Their function was to protect a centrally located troop concentration and maneuver area within the Polish salient. Behind these fortresses, another network covered key concentration and transit nodes at Brest-Litovsk, Grodno, and Kovno. Lesser fortifications along the Bug-Narew line defended against incursion from East Prussia. However, technological advances, including more powerful high explosives

[11] RGVIA, f. 402 (*Komitet po mobilizatsii voisk Glavnogo shtaba*), op. 1, d. 833, ll. 1–4, and f. 400, op. 4, d. 54, ll. 6–60b.

[12] On the time required, see, RGVIA, f. 2000 (*Glavnoe upravlenie General'nogo shtaba*), op. 1, d. 1812, ll. 67–8.

[13] S. K. Dobrorol'skii, *Mobilizatsiia russkoi armii v 1914 godu* (Moscow, 1929), 110.

and ever heavier siege artillery, subjected these and other fortifications to a continuing race with obsolescence. By the beginning of the twentieth century, the Russians had already sunk more than a half-billion rubles of military capital into the Polish salient. Recent advances in siege artillery promised still another round of expenditures for fortress modernization.[14]

Expansion of the strategic railroad net in the West constituted another formidable challenge. After a burst of construction during the 1890s, only one additional double-track line – the only kind of genuine utility for military traffic – was completed in 1907 to link Siedlce in the Polish salient with Bologoe and other points in central Russia. This line plus other refinements yielded a 1914 total of six double-track lines and two single-track lines feeding into the forward theater. Their collective throughput capacity was slightly more than 200 military trains per day (transiting a corps-size formation required 135).[15] In contrast, Germany and Austria-Hungary together possessed double the Russian railroad throughput capacity to the inter-state frontier. For the Russians, full transit to concentration and deployment required a total figure in excess of 4,000 troop trains, with the greatest demand for throughput capacity occurring during the second and third weeks of mobilization and concentration.[16] Double-track connecting lines from key population nodes in the Russian interior were few, and north-south connecting lines between the Pripet and the Dnepr were non-existent. Their absence precluded the ability to shift troops during the final stages of transit to the West from one major strategic axis to another. In the end, the brittleness of the rail transit system was perhaps surpassed only by the brittleness of the complex troop mobilization system.[17]

These and related structural rigidities set the terms for Russian war planning. Thanks to a large population and the fruits of conscription, Russia in 1914 might field as many as 37 army corps, each consisting of

[14] For a survey, see, [F. F. Palitsyn and M. V. Alekseev] *Doklad o meropriiatiiakh po oborone Gosudarstva, podlezhashchikh osushchestvleniiu v blizhaishee desiatiletie* [*v. sekretno*] (St. Petersburg, n. d.), 53–8.

[15] In the parlance of the day, troop trains were called echelons, and besides a locomotive, a tender, and six passenger cars, they numbered exactly 44 two-axle freight and platform cars; see, Golovin, *Plan voiny*, 10–11.

[16] K. P. Ushakov, *Podgotovka voennykh soobshchenii Rossii k mirovoi voine* (Moscow-Leningrad, 1928), 58–9.

[17] For the broader allied military perspective, see, D. N. Collins, "The Franco-Russian Alliance and Russian Railways, 1891–1914," *The Historical Journal*, XVI, no. 4 (December 1973), especially 785–88.

about 30,000 troops in a large infantry-heavy combined arms formation. Of these corps fewer than 30 were immediately available for deployment to the West, and of this number, 25 might initially concentrate opposite Germany and Austria-Hungary.[18] Between 1873 and 1914, the painful requirement to attain mass over time with limited assets and means against two powerful adversaries loomed over everything, including the development of strategies, plans, and courses of action. Both primary systems that underlay pre-1914 Russian preparation for war accepted the necessity for an initial defensive phase, if only to cover mobilization and concentration, followed by transition to the offensive. The historian I. I. Rostunov once observed that all Russian plans for war were ultimately offensive; the only questions were when, and with what means?[19] In pursuit of palatable answers to these two questions, Russian planners sought every possible advantage from a situation fraught with practical limitations. More often than not, General Staff officers, or *genshtabisty* as they were called, discovered that only incremental advantage was attainable, and then only at the margins, and ultimately only at the expense of risk. They also discovered that tinkering with any major element within the calculus was to court exceptional risk. In any case, risk required either mitigation or rationalization – or perhaps a generous portion of each – with an admixture of resignation and old-fashioned Russian fatalism.[20]

The Miliutin-Obruchev System

It was the requirement to balance risk with mitigation that produced the Miliutin-Obruchev system. Named for the reform-minded minister of war and his primary planning assistant, General N. N. Obruchev, who served as chief of the Main Staff between 1881 and 1897, this system persisted for nearly three decades, from the 1880s until 1910. The strategy inherent in the Miliutin-Obruchev system was decidedly defensive-counteroffensive. Because an underdeveloped strategic railroad network precluded the rapid attainment of mass in the West, the governing idea was to defend forward during the initial period of a general European war.

[18] There were 27 corps in European Russia, plus three in the Caucasus, two in Turkestan, and five in Siberia; see, A. F. Rediger, *Komplektovanie i ustroistvo vooruzhennoi sily*, 2 vols. (St. Petersburg, 1913), II, 161.

[19] I. I. Rostunov, *Russkii front pervoi mirovoi voiny* (Moscow, 1976), 95.

[20] In *Strategy and Power in Russia*, 451, Fuller views the equation in less positive terms, arguing that "Russian strategy for war in 1914 was still dependent on magic and miracle."

Defense of imperial territory mandated the retention of major fortresses within the Polish salient and on its shoulders. Their garrisons, together with active duty formations deployed forward during peacetime within the three frontier military districts, were initially to defend and to function as covering forces for the more slowly mobilizing and transiting reinforcements from the interior military districts. The network of fortresses assured interior lines of operation for Russian troop formations within the Polish salient. Once the mobilization and rail transit of troops from the Russian interior to forward concentration had wound their lengthy course, various major formations might deploy to assume the counteroffensive. In other words, the defensive-offensive strategy inherent in the Miliutin-Obruchev system stipulated a follow-on course of action.[21]

However, simultaneous offensive operations against both Germany and Austria-Hungary remained out of the question. The concept was to defend against Germany while conducting offensive operations deep into Habsburg lands, indeed, even as far as Vienna and Budapest.[22] Not until 1912 did alliance-inspired staff talks with the French refer to a timetable for Russian offensive operations against Germany. Meanwhile, as palliative for an inadequate railroad network, Russian peacetime deployments tilted inexorably toward the western military frontier. Soon, the majority of the peacetime active army was deployed either within the Polish salient or close to its shoulders. Another palliative, one born of desperation, was to plan massive cross-border cavalry raids at the outset of hostilities with as many as 200 squadrons into East Prussia and Galicia. The purpose of these raids would be to disrupt enemy railroad networks in order to buy time for the more slowly mobilizing and deploying Russian army.[23]

Although the Miliutin-Obruchev system persisted – on paper at least – until 1910, Mobilization Schedule No. 18 (1 January 1903) probably marked its apogee. In the event of war in the West, this schedule would have mobilized 1,472 infantry battalions, 1,035 cavalry squadrons, and 4,558 guns. These considerable forces were to be deployed in seven

[21] Codified by V. N. Domanevskii in V. M. Alekseeva-Borel', *Sorok let v riadakh russkoi imperatorskoi armii: General M. V. Alekseev* (St. Petersburg, 2000), 230–31; see also, David Alan Rich, *The Tsar's Colonels: Professionalism, Strategy, and Subversion in Late Imperial Russia* (Cambridge, MA, 1998), 162–71.

[22] Zaionchkovskii, *Podgotovka Rossii k imperialisticheskoi voine*, 44–6, 65–7; see also, S. K. Dobrorol'skii, "Strategicheskie plany storon k nachalu mirovoi voiny," *Voennyi sbornik* [Belgrade], no. 2 (1922), 53–5.

[23] RGVIA, f. 400, op. 4, d. 50, ll. 20–20ob.

armies, two of which would secure the Baltic coast and St. Petersburg, while the remaining five would be arrayed on the borders with Germany and Austria-Hungary. In a calculus that would remain much the same until 1914, the common Russian assumption was that the Germans might launch an offensive by the twelfth day of mobilization and Austria-Hungary by the sixteenth day. During this span, War Minister A. N. Kuropatkin lamented that "the concentration of our field forces would not have been completed and our second-line reserve troops and Cossacks would not even have begun transit." His conclusion was that "this time lag inescapably subordinates our initial form of strategic troop deployments to the idea of defense." However, Kuropatkin remained convinced that a counteroffensive against Austria-Hungary was "the general idea of our action against the enemy allies," but only after full Russian reinforcements had arrived in theater, roughly 30–45 days after the onset of mobilization.

A major problem inherent in this and preceding schedules for war in the West was that the lopsided deployments inherent in the Miliutin-Obruchev system rendered non-western contingencies a grave problem. With their focus on a possible war against Germany and Austria-Hungary, Miliutin and Obruchev had anticipated neither a war with Turkey in 1877–78 nor the antagonisms in Central Asia and the Far East that came later.[24] Moreover, between 1885 and 1897, a joint army-navy expedition to seize the upper Bosporus remained an *idée fixe* for Russian war planners.[25]

One solution to imperial overreach was to rely on the Russian navy as the first line of defense, especially on the Black Sea and in the Far East. However, just like the army, the navy suffered from its own version of multiple potential adversaries along multiple axes. The Baltic littoral and the imperial capital at St. Petersburg were vulnerable to naval and amphibious assault from several quarters, while Anglo-Russian contention at the Turkish Straits and during the "Great Game" in Central Asia made the British navy a force with which to contend. The Russian response during the last quarter of the nineteenth century was to construct Europe's third largest navy, but even its numerous assets were

[24] P. A. Zaionchkovskii, *Voennye reformy 1860–1870 godov v Rossii* (Moscow, 1952), 287–88.

[25] O. R. Airapetov, *Zabytaia kar'era "Russkogo Mol'tke": Nikolai Nikolaevich Obruchev (1830–1904)* (St. Petersburg, 1998), 274–87.

inadequate to address all possible combinations for potential combat in various unlinked theaters of war.[26] Worse, from the perspective of the War Ministry, the navy became a serious contender for resources that might be allocated for other purposes, especially fortress modernization.

Still worse, the War Ministry's traditional emphasis on the western military frontier did not completely correspond with the tsar's strategic priorities. General Kuropatkin and like-minded *genshtabisty* had long held the conviction that the empire's fate in any future general war would be determined in the West. However, Nicholas II did not fully share this view during the first decade of his reign. For various reasons, he was intent on Far Eastern expansion, and he therefore froze military preparations in the West and diverted key military and naval resources to the Far East. By 1903, plans were shelved for a possible amphibious expedition to seize the Bosporus.[27] More significantly, during 1902–03, the tsar ordered the Main Staff, Russia's pre-1905 version of a capital staff, to consider withdrawing peacetime troop deployments in the West farther into the Russian interior to free forces and resources for commitment against Japan and China.[28] With no standing institutional or advisory brake on autocratic power or preference, the natural result was strategic over-extension. Fortresses and railroads in the West languished while scarce funds flowed into construction of the Trans-Siberian Railroad and pre-dreadnought battleships for the Russian Pacific Squadron.[29]

On the periphery, just as in the West, various cumbersome troop mobilization arrangements dictated a defensive-counteroffensive strategy. Against Japan, for example, Kuropatkin's strategy was initially to defend behind a naval and ground force screen, and then to mobilize reinforcements in depth, including at least two corps from European Russia, to provide sufficient mass for counteroffensive operations to push invading Japanese troops from the Asian mainland. Lack of local infrastructure in Manchuria and the low carrying capacity of the Trans-Siberian Railroad and its branches dictated a prolonged defensive phase before transition

[26] M. A. Petrov, *Podgotovka Rossii k mirovoi voine na more* (Moscow-Leningrad, 1926), 8–9, 22–4, 30–4, 47–52.

[27] Ibid., 86–7.

[28] RGVIA, f. 400, op. 4, d. 433, ll. 37–8.

[29] Bruce W. Menning, "Neither Mahan nor Moltke: Strategy in the Russo-Japanese War," in John W. Steinberg, *et al.* (eds.), *The Russo-Japanese War in Global Perspective: World War Zero*, 2 vols. (Leiden, 2005, 2007), I, 133–34, 139–40.

to the counteroffensive.[30] The same circumstances dictated close cooperation between the army and navy to buy time for a buildup, but not even the mid-1903 appointment of a Far Eastern viceroy at Port Arthur assured such cooperation in the absence of effective higher-level coordination and clearly established priorities. In distant St. Petersburg, the tsar presided only haphazardly over a small maze of competing ministries, conferences, and special committees whose task it was to fashion a consistent Far Eastern policy. Interest groups and favorites figured far too prominently in decision-making, and the tsar as autocratic arbiter was not above retaining mastery by playing personalities and factions against one another.[31]

Imperfect coordination led to poor performance and baleful consequences. When war came unexpectedly to the Far East in early 1904, the flow of events rapidly overwhelmed sketchy plans and scarce resources. A surprise Japanese naval attack crippled the Russian Pacific Squadron at Port Arthur, and Japanese armies landed in Manchuria to score a series of successes against ineptly employed and poorly led Russian ground force screens. Although Kuropatkin had originally anticipated sending only two corps from European Russia to the Far East, more than nine partial troop mobilizations eventually dispatched eight corps eastward from European Russia to the battlefields of Manchuria. Once concentrated, these forces fared no better than the earlier screening detachments. By the time the Russo-Japanese War ended in September 1905 (N. S.), it had bankrupted the treasury and made a shambles of Russian strategic deployments in the West. The credibility of the Franco-Russian alliance was badly frayed, the empire was wracked by revolution, all classes of military supplies were depleted, and three Russian naval squadrons had ceased to exist.[32]

Post-1905 Dysfunctions and Reassessments

Russia had fewer than nine years to overhaul its army and resurrect its navy before the outbreak of a general European war. Three factors figured prominently in the process. One was the uphill fight for domestic recovery and maintenance of the fragile domestic peace. The second was the impact

[30] Iu. F. Subbotin, "A. N. Kuropatkin i Dal'nevostochnyi konflikt," in I. S. Rybachenok, L. G. Zakharova, and A. V. Ignat'ev (eds.), *Rossiia: Mezhdunarodnoe polozhenie i voennyi potentsial v seredine XIX-nachale XX veka* (Moscow, 2003), 161–62.

[31] Menning, "Neither Mahan nor Moltke," 142–44.

[32] On the consequences of defeat, see, K. F. Shatsillo, *Ot Portsmutskogo mira k pervoi mirovoi voine* (Moscow, 2000), 14–25.

of Russia's changed geopolitical circumstances on threat assessment and corresponding defense requirements. The third was constrained resources. The Russo-Japanese War had cost Russia 6.119 billion rubles, including 3.943 billion for debt service.[33] It was not until 1909 that the economy would recover sufficiently to obviate the necessity for foreign loans. Low-cost military change might come immediately; more expensive reforms and programs came more slowly during the last five years before the outset of hostilities in 1914.

In contrast with the period before the Russo-Japanese War, the post-1905 institutional context was different, in part because of the domestic political settlement and in part because of new and untried mechanisms for overall strategic direction and planning. In accordance with the October Manifesto of 1905 and the Fundamental Laws of 1906, the tsar now had to contend with the Duma, an elected legislative assembly. Although the Council of Ministers that constituted the tsar's cabinet was responsible to the throne and not the Duma, the latter retained the right – with limitations – to appropriate monies and to subject ministers to interpellation. Meanwhile, the tsar retained an absolute constitutional monopoly over foreign and military affairs.[34] He was often indecisive, always loathe to deal directly with the Duma, and usually not above playing ministries and constituencies against one another in a reversion to his pre-1905 practice of "divide and misrule."

Not surprisingly, major components within the new governing system failed to mesh until 1907, when changed electoral laws produced a reasonably compliant Duma, and when the appointment of P. A. Stolypin produced a prime minister in whom the tsar had confidence. However, even Stolypin did not always see eye-to-eye with the tsar on the nature of the new constitutional order. Until his assassination in 1911, the prime minister remained intent on creating a united government that reported through him to the tsar and that worked in a coordinated and productive manner with the Duma. A major objective was to prevent the kind of imperial overreach and inter-ministerial wrangling that had figured so prominently in the events and decisions leading to Far Eastern defeat. In contrast, the tsar remained intent on "roll back" to preserve and even

33 Boris Ananich, "Financing the War," in Steinberg *et al.* (eds.), *The Russo-Japanese War in Global Perspective*, I, 463.

34 David MacLaren McDonald, *United Government and Foreign Policy in Russia, 1900–1914* (Cambridge, MA, 1992), 87–92, 97–102; the more strictly foreign policy dimension is outlined in A. V. Ignat'ev, *Vneshniaia politika Rossii, 1907–1914: Tendentsii. Liudi. Sobytiia* (Moscow, 2000), 26–42.

expand traditional autocratic prerogatives.[35] With these and other crosscurrents at work, it would require two additional years and several serious embarrassments before Stolypin might begin to preside over anything that resembled a reasonably harmonious whole.

Other crosscurrents swirled among lesser institutions and actors whose function it was to facilitate collaboration and cooperation in planning for and allocating resources to imperial defense. In 1905 and 1906, the first wave of defeat-inspired military and naval reform witnessed the creation of the Main Directorate of the General Staff (*GUGSh*) and the Naval General Staff, each of which was intended to become its service's primary planning focus for future war. *GUGSh* was torn from the side of the Main Staff, in large part to centralize that organization's disparate and ill-coordinated planning agencies. As was the case with the Great German General Staff, the *GUGSh* chief, General F. F. Palitsyn, reported directly to the throne, bypassing General A. F. Rediger's War Ministry, which now became an organ for administration and support. Despite a complement of 200 officers and civil servants and an impressive mandate that embraced nearly all aspects of ground force preparation for war, *GUGSh* was slow to fulfill its promise. Until 1909, the troop mobilization committee still resided with the Main Staff, and General Palitysn exercised no authority over either the War Ministry, the all-powerful chiefs of the military districts, or the several grand dukes who served as inspectors-general for the more important service branches.[36] For its part, the more modest Naval General Staff (16 officers) remained subservient to its parent Naval Ministry, with the result that planners were initially subsumed into the post-1905 fight for additional resources to rationalize capital-ship construction for the war-ravaged navy.[37]

The task of another newly created institution, the State Defense Council under the chairmanship of the Grand Duke Nicholas Nikolaevich, was to impose unity on diverse agencies and programs in defense of the realm. However, neither the grand duke nor the Council's membership,

[35] David Alan Rich, "Russia," in Richard F. Hamilton and Holger H. Herwig (eds.), *The Origins of World War I* (Cambridge, UK, 2003), 196–97; see also, McDonald, *United Government*, 97–102.

[36] A. S. Skvortsov, A. S. Rushin, et al. (eds.), *General'nyi shtab Rossiiskoi armii: Istoriia i sovremennost'* (Moscow, 2006), 71–2, 77–9; and Galina Kozhevnikova, *Glavnoe upravlenie General'nogo shtaba nakanune pervoi mirovoi voiny (1910–1914)* (Moscow, 1998), 10–11, 113–14.

[37] Evgenii F. Podsoblyaev, "The Russian Naval General Staff and the Evolution of Naval Policy, 1905–1914," *The Journal of Military History*, LXVI, no. 1 (January 2002), 42–3.

which lopsidedly favored the army, sympathized with the tsar's navalist pretensions. A tactician and not a strategist, the grand duke cared little for administrative detail or for the rough-and-tumble of Duma-style politics. In the end, his devotion to the army strained his relationship with the tsar, with the result that the grand duke's influence rose and fell, often in response to crisis-driven politics of the moment. Nonetheless, during the entire 1905–14 period, Grand Duke Nicholas Nikolaevich remained commander of the Guards and chief of the St. Petersburg military district. Differences with his imperial cousin notwithstanding, access to the tsar meant that the grand duke continued to wield considerable influence over the promotion and assignment of senior army officers.[38]

Against the background of fragmented institutions and powerful but contentious personalities, threat assessment became an important weapon in the fight to galvanize programs, garner resources, and unify action. While the Naval General Staff initially spun its paddles in working out a strategy that might justify anticipated dreadnought construction, officers in *GUGSh* operated under their own set of constraints. Like the navy, the army was starved for resources, and many active troop formations were scattered across the land, putting down post-1905 urban and rural disturbances. The recent war had exhausted all categories of wartime stocks, and railroad assets were either depleted or dispersed across the network linking the Far East with European Russia.[39] At the same time, the pendulum of preparedness for anticipated conflict was decisively swinging back to an undiluted emphasis on the western military frontier. Repeated tsarist guidance during late 1906 enjoined *GUGSh* to base its net threat assessment on the "worst case," that is, on war against the Triplice in which Germany initially deployed the preponderance of its forces against Russia.[40]

In light of constraints and guidance, threat assessments and the recommendations flowing from them were both pessimistic and cautious. They would remain so for the next five years. Thus, the two *genshtabisty*

[38] L. G. Beskrovnyi, *Armiia i flot Rossii v nachale XX v.* (Moscow, 1986), 49–50, 64–6; on the grand duke, see, P. A. Zaionchkovskii, "Vysshee voennoe upravlenie. Imperator i tsarstvuiushchii dom," in L. G. Zakharova, Iu. S. Kukushkin, and T. Emmons (eds.), *P. A. Zaionchkovskii 1904–1983 gg. Stat'i, publikatsii i vospominaniia o nem* (Moscow, 1998), 93–5.

[39] K. F. Shatsillo, *Rossiia pered mirovoi voine* (Moscow, 1974), 15; see also, Aleksandr Rediger, *Istoriia moei zhizni*, 2 vols. (Moscow, 1999), II, 9–10, 81–2.

[40] RGVIA, f. 1759 (*Shtab Kievskogo voennogo okruga*), op. 1, d. 3539, l. 40; and f. 2000, op. 1, d. 492, ll. 67–9.

compiling the initial post-1905 assessment, General M. V. Alekseev and Colonel S. K. Dobrorol'skii, labeled Germany as the primary objective, since that state was the "heart and soul" of the hostile coalition. They conceded that Russia could not contend with simultaneous hostilities in the East and West, and therefore acknowledged that only skillful diplomacy might forestall any additional deterioration in relations with Japan and China. They also conceded that Mobilization Schedule No. 18 was no longer viable, and even asserted that it existed "only on paper" in view of Russia's lack of military preparedness. Until the advent of reform and additional resources, they recommended that Russian strategic deployments in the West be withdrawn farther into the interior to preclude envelopment within the Polish salient at the outset of possible war.

Nonetheless, they did not relinquish the defensive-counteroffensive strategy inherent in the Miliutin-Obruchev system. After the outset of war, once sufficient mass was built up, they anticipated transition to offensive operations, but with Germany now the primary objective. More ominously, they also conjectured that under special circumstances Russian forces might engage in offensive operations before the completion of mobilization and concentration.[41] Two years would elapse before *GUGSh* seriously revisited this assessment, making it more comprehensive and linking it with a reform program for improved capabilities and readiness.

With no viable interim alternative on which to base planning for war, *GUGSh* officers reverted to a mildly revised scheme, Mobilization Schedule No. 18 (Restored). In accordance with tsarist guidance, its dispositions were heavily weighted against Germany. Altogether, the scheme provided for the deployment in the West of eight armies, including a Vistula Army concentrated well forward in the Polish salient. The rear and flanks of this army were covered – at least on paper – by three additional armies.[42]

There were, however, several glaring problems: the network of protective fortresses was now obsolescent, and the lone army deployed forward on the Vistula remained beyond the range of timely reinforcement by the three nearest armies until the twentieth day of mobilization, while the Germans might attack in force as early as the twelfth day. Worse, various organizational deficiencies and supply shortages meant that five

[41] Ibid, f. 2000, d. 97, ll. 4, 9, 10ob., 18–21, 24–8ob., 31–4.

[42] Ibid., d. 492, ll. 14–27; see also, the commentary in Zaionchkovskii, *Podgotovka Rossii k imperialisticheskoi voine*, 156–58; and in Rostunov, *Russkii front*, 90–2.

of the eight armies were not combat ready. Nevertheless, the restored schedule went into effect in late 1907, perhaps because of a lack of time and resources to develop a better alternative, or more possibly because planners anticipated that a combination of diplomatic initiatives, new resources, and the beginning of a fresh five-year budget cycle in 1908 would soon give rise to an entirely new schedule.[43]

As officers within *GUGSh* and the War Ministry wrestled with the problem of combat readiness, Foreign Minister A. P. Izvol'skii's initial emphasis was on the recovery of credibility and prestige lost during the Russo-Japanese War, while not denying opportunities to attain traditional Russian foreign policy objectives in the Balkans and at the Turkish Straits. Thus, just as in the Alekseev-Dobrorol'skii assessment, the focus of Russian foreign policy reverted to the West. In February 1907, the foreign minister pointedly noted that "the center of gravity of our influence must not be on the Far East, but on the west..."[44] That same month, Izvol'skii clearly enunciated Russian strategic priorities, observing that "Russia might lose her maritime provinces [in the Far East], but if she loses the ability to determine outcomes in the Balkans, she might be stricken from the ranks of the Great Powers."[45]

Izvol'skii's assertions represented only a general consensus within tsarist governing and military circles over larger foreign policy goals. In contrast, there was at best only partial consensus over the ways and means to pursue them, along with lesser objectives. The result was a mixture of limited progress, contradiction, and even embarrassment. Rapprochement with Japan and England during 1907 reduced tensions and competition in both the Far East and Persia. Still, Russian aspirations at the Straits remained a possible bone of contention with both France and England. There was acknowledgement that a revitalized alliance with France remained the cornerstone of Russian foreign policy in Europe. Yet, at the same time, the tsar and Izvol'skii chose to pursue a balancing policy between Germany and England – even as England was drawing closer to France, thanks in part to the impact and implications of the First Moroccan Crisis during 1905–06.[46]

[43] RGVIA, f. 2000, op. 1, d. 508, ll. 69–70.

[44] Quoted in A. A. Polivanov, *Iz dnevnikov i vospominanii po dolzhnosti voennogo ministra i ego pomoshchnika, 1907–1916 g.*, ed. A. M. Zaionchkovskii (Moscow, 1924), 18.

[45] Quoted in [Palitsyn and Alekseev] *Doklad*, 10; see also, Shatsillo, *Ot Portsmutskogo mira*, 33–42.

[46] Ignat'ev, *Vneshniaia politika Rossii*, 23, 35–8.

These and other anomalies fed internal differences and dysfunction at the highest reaches of the Russian government. Grand Duke Nicholas Nikolaevich's State Defense Council did not share the tsar's emphasis on creation of a dreadnought-style navy either to serve as a balancing instrument between England and Germany or to satisfy cravings for imperial prestige. However, for various reasons, the Council of Ministers and significant elements within the Duma were sympathetic to naval construction programs, even during a period of financial stringency. In early 1907, when the State Defense Council failed to agree with the Naval Ministry's proposal to initiate a new program for capital ship construction, the tsar overrode counsel to approve a limited program, including completion of four capital ships for the Baltic Fleet.[47] Both the grand duke and his State Defense Council immediately lost credibility and influence, and along with them, much of the capacity to harmonize policies with strategies and acquisitions.

With Stolypin's version of a united government still unrealized, and with the State Defense Council's fortunes already on the wane by mid-1907, there was precious little coordination of security affairs and no consistent institutional brake on a reversion to adventurism in foreign policy. On the basis of only limited support from the tsar and General Palitsyn, Foreign Minister Izvol'skii proposed in late 1907 to mount a preemptive war against Turkey to seize control of the Straits. However, fear of Great Power intervention and the Russian armed forces' appalling lack of readiness soon doomed the project. Neither Prime Minister Stolypin nor Finance Minister V. N. Kokovtsov was originally privy to the scheme, and when word leaked out, they were furious because the prospect of war over the Straits would imperil the fragile domestic peace. The forum for airing and resolving differences was not the State Defense Council, but a time-worn mechanism, the inter-ministerial special conference. Before cooler heads prevailed, Stolypin threatened resignation, and both Foreign Minister Izvol'skii and *GUGSh* chief Palitysn suffered embarrassment and loss of credibility.[48]

On a short tether and with a less than predictable life expectancy, the State Defense Council successfully catalyzed and coordinated only the most fundamental and lowest cost military reforms. These included

[47] For the most comprehensive account, see, Shatsillo, *Ot Portsmuskogo mira*, 92–103; see also, Podsoblyaev, "The Russian Naval General Staff," 43–4.

[48] McDonald, *United Government*, 112–19; and Shatsillo, *Ot Portsmutskogo mira*, 37–51.

a reduction of active service terms for conscripted soldiers and sailors, improvements in living conditions for troops, and modestly increased salaries and allowances for officers.[49] More significantly, War Minister Rediger and the grand duke received the tsar's blessing to institute the Supreme Certification Commission, a mechanism for imposing rigor on officer promotion and assignment. Under the Commission's aegis, the army officer corps underwent a veritable purge, with nearly one-fifth of its complement retired or released from active service. There was, however, at least one major impediment: the tsar often imposed his own criteria, including political reliability and personal acquaintance, on senior-level command appointments to corps and military districts. A lamentable consequence was that officers with recent combat experience in Manchuria were less than adequately represented at the upper reaches of the command and staff structure.[50]

Incremental and imperfect as these changes were, they contrasted sharply with the larger picture of immobility and disarray. With the Baltic Fleet badly in need of lesser vessels for coastal defense, Duma deputies friendly to the army questioned allocations for capital ships. Finance Minister Kokovtsov limited the army to bare-bones continuation budgets, and for a time he and Stolypin held additional naval appropriations hostage to organizational reform. Outside official channels a small group of army officers labeled "the Young Turks" dared to engage in regular consultation with members of the Duma. Inside the military establishment, *GUGSh* chief Palitysn shared details of the war planning process not with the War Ministry, but with the chiefs of the military districts.

Meanwhile, War Minister Rediger knew nothing of military obligations inherent in the secret alliance with the French. For their part, the French might only guess at the true state of the Russian army's readiness for war. Nicholas II and not Prime Minister Stolypin remained the supreme arbiter for Russian foreign and military policy, but the tsar often kept ministries in the dark or played them against each another to retain

[49] A. G. Kavtaradze, "Military Reforms of 1905–12," *Great Soviet Encyclopedia*, tr. 3rd ed., 31 vols. and supplement (New York, 1973–1983), V, 278–79.

[50] P. A. Zaionchkovskii, "Russkii ofitserskii korpus nakanune Pervoi mirovoi voiny," in Zakharova, Kukushkin, and Emmons (eds.), *P. A. Zaionchkovskii 1904–1983 gg.*, 37–40; on the selection rates for combat veterans, see, John W. Steinberg, "The Quest to Reform: The Education, Training, and Performance of the Imperial Russian General Staff, 1898–1914," unpublished MS, 68–9.

ultimate authority and freedom of political maneuver. At the same time, the ministers themselves were sometimes less than forthcoming in their reports to the tsar.[51]

The result was disaster in foreign affairs. Frustrated by the failure of his gambit for outright seizure of the Straits, Izvol'skii resorted to an indirect approach through Austria-Hungary. Without consulting the tsar, Izvol'skii apparently agreed with the Austro-Hungarian foreign minister, Count Alois Aehrenthal, that Vienna might formally annex Bosnia and Herzegovina, both nominally belonging to the Ottomans, in exchange for Vienna's support for an altered regime at the Straits to permit free passage for Russian warships (during the Russo-Japanese War, six Russian battleships had remained idle on the Black Sea). The arrangements at both the Straits and within the two Balkan provinces had resulted from international agreements dating to the 1870s and 1880s. However, Austria-Hungary had *de facto* occupied Bosnia and Herzegovina since 1878, while Russian claims at the Straits rested on mere assertion.

Unknown either to Izvol'skii or the tsar, Russia during the 1880s had secretly affirmed the Dual Monarchy's claim to the provinces, a fact that Aehrenthal leaked to the European press. Therefore, in October 1908 (N. S.), when Emperor Franz Joseph abruptly informed Nicholas II that annexation was a *fait accompli*, the duped and apparently hypocritical Izvol'skii was left shuttling in vain among the European powers to mobilize support against Vienna. In March 1909, with the threat of war between the Romanovs and Habsburgs hanging in the air, Germany sent St. Petersburg a near-ultimatum hinting at the possibility of a wider coalition war unless Russia formally recognized the annexation.[52] The *coup de grace* came at a meeting of the Council of Ministers on 6 March 1909, when War Minister Rediger abjectly admitted that the Russian army was in no condition to countenance war over the Bosnian annexation crisis. Russian humiliation was complete: France had offered no assistance, the Serbian nationalists who counted the provinces theirs felt betrayed, and Russia was linked in perfidy with Austria-Hungary. Izvol'skii's gambit

[51] Peter Gatrell, *Government, Industry and Rearmament in Russia, 1900–1914: The Last Argument of Tsarism* (Cambridge, UK, 1994), 125–52; V. A. Sukhomlinov, *Vospominaniia* (Berlin, 1924), 235; A. V. Ignat'ev, "P. A. Stolypin i vossozdanie voenno-morskogo flota posle russko-iaponskoi voiny," in Rybachenok, Zakharova, and Ignat'ev (eds.), *Rossiia*, 182–88; Rediger, *Istoriia moei zhizni*, II, 11–12, 226–27.

[52] For an overview, see, Oleg Airapetov, *Vneshniaia politika Rossiiskoi imperii (1801–1914)* (Moscow, 2006), 526–32.

had brought severe embarrassment to Imperial Russia, "a diplomatic Tsushima," as one Duma deputy put it.[53]

The Bosnian crisis lent additional urgency to the impulse for military renewal that antedated the crisis itself. On 2 March 1908, the tsar had mandated unity in planning, commenting that "the general plan for state defense must be brief and clear – for one or two decades" and that "we must unswervingly bring it to fruition."[54] However, this plan would evolve under the auspices of the Council of Ministers, not the State Defense Council. In April, the grand duke requested reassignment and recommended that the State Defense Council be abolished.[55] In June, Rediger's War Ministry at last received supplementary funds of 293 million rubles spread over seven years for limited new armaments and for acquisitions to make good on losses suffered in 1904–05.[56] A year later, in the wake of the Bosnian crisis and another threatened resignation, Stolypin would come to wield enhanced power over a ministerial system that began to approach his ideal of a united government. Although the tsar still retained absolute authority over military and foreign policy, the appointment in June 1909 of Stolypin's brother-in-law, S. D. Sazonov, as assistant foreign minister promised to reduce inter-ministerial dissonance and bring greater harmony to the coordinated pursuit of internal and external affairs. In September 1910, Sazonov would become foreign minister in his own right. However, the war and navy ministries remained beyond Stolypin's direct reach.[57]

Meanwhile, even as the Bosnian crisis was in its early stages, General Alekseev during mid-1908 had returned to threat assessment, but with important departures from the previous iteration. His revamped assessment came after rapprochement with Japan and Britain, but in light of a sharpening struggle for resources. Acuity of need in no small part accounted for the new report's alarmist tone, highlighted by the starkness of the threat picture, and reinforced by the supplanting of Colonel Dobrorol'skii with *GUGSh* Chief Palitsyn as co-author.[58] The new Alekseev-Palitsyn variant on threat assessment also unabashedly

[53] Rediger, *Istoriia moei zhizni*, II, 277.
[54] Quoted in Polivanov, *Iz dnevnikov*, 43.
[55] Ibid., 43, 45, 48–9.
[56] Shatsillo, *Ot Portsmutskogo mira*, 112.
[57] David McDonald, "Tsushima's Echoes: Asian Defeat and Tsarist Foreign Policy," in Steinberg, *et al.* (eds.), *The Russo-Japanese War in Global Perspective*, I, 551–57.
[58] The original report is at RGVIA, f. 2000, op. 1, d. 156; the printed version is [Palitsyn and Alekseev] *Doklad*.

advocated a reform agenda. Distinctions aside, there were major continuities between Alekseev-Dobrorol'skii and Alekseev-Palitsyn. The latter held that the primary threat remained the same: the Triple Alliance. The threat in the Far East – Japan, and, to a lesser extent China – required vigilance and enhanced peacetime military preparations, but the western frontier clearly assumed uncontested priority for potential wartime commitments.

The two generals conceded that it would be ideal to field sufficient forces to conduct successful simultaneous wars in both the East and West, but unfortunately "these considerations could attain practical realization only in the very distant future." Therefore, in the event of simultaneous hostilities in the East and West, the Russian Far East might not expect reinforcement from the imperial interior until perhaps the final period of conflict in the West. Considerations of time alone underlay these assertions. More than 20 corps from European Russia might be mobilized and transited to the western military frontier in two months, while the deployment of 10 corps from European Russia to the Far East required something in excess of six months.[59]

The same degree of certainty characterized other assessments in the Alekseev-Palitsyn report. Absent England from the threat calculus (except at the Straits), the Triple Alliance occupied center stage, but lesser potential adversaries were in the wings. Turkey and Romania (and to some extent, Sweden) now figured at least informally among Russia's coalition adversaries, so Alekseev and Palitsyn generously added their armed forces to the long list of Russia's potential wartime foes in the West. The resulting aggregate of 2,133.5 hostile battalions outnumbered the Russians by nearly 600 battalions, even if *GUGSh* factored into the equation reinforcements from the Caucasus and Turkestan. Even with England out of the picture, the odds against Russia were stupendous. And, even if Germany in a future European war turned initially against France, the resulting force correlations along Russia's western military frontier were only equal.[60]

Against these odds, Alekseev and Palitsyn emphasized the importance of defense and depth. In one of the first formal departures from the Miliutin-Obruchev system, they concluded that "*we must concentrate our main forces outside the forward theater* [italics in original]."[61] Because

[59] [Palitsyn and Alekseev] *Doklad*, 9, 11; cf., Fuller, *Strategy and Power in Russia*, 423–26.
[60] [Palitsyn and Alekseev] *Doklad*, 12–17.
[61] Ibid., 48.

Germany and Austria-Hungary still retained a two-to-one superiority over Russian railroad transit capacity to the state frontier, there was simply no other military answer to the time-space-mass dilemma. Only as an exception the authors held that "special circumstances would be required for even a numerically superior Russian army to count on beginning a war not on its own territory, but beyond the borders of its own homeland with a rapid transition to the offensive following completion of deployment." But, they added, "such action will long be impeded by our distances, and even more importantly, by the absence of a sufficient number of railroads."[62]

Consequently, the two officers would anchor Russian defensive dispositions for the potential Northwest Front on the middle flow of the Niemen and for the Southwest Front on key terrain not within the Polish salient, but in the vicinity of Starokonstantinovo, west of the railroad paralleling the Austro-Hungarian frontier between Berdichev and Shepetovka. The assessment even envisioned a supplementary second defensive line deeper in the interior, along the Northern Dvina and the Dnepr. However, Alekseev and Palitsyn did not advocate complete abandonment of the forward theater. Instead, they recommended the selective modernization and creation of fortresses within Russian Poland and on its shoulders, both to offer initial resistance to invaders and to assure freedom of maneuver for advancing Russian forces that were now to be initially deployed at a greater distance from the state frontier.[63]

Other important recommendations involved changes to peacetime deployments, along with a coherent program for army reform and rearmament. Redeployment and reform would address a number of substantial distortions that had crept into the Miliutin-Obruchev system over the previous three decades. Among these were important and expensive discrepancies between territorial recruitment in the Russian interior and forward deployment within the Polish salient, the demonstrated inefficiency of standing cadre-based reserve formations, the debilitating multiplication of troop types, and the longer-term problem of western-oriented peacetime dispositions in the event of a possible war in the Far East. An additional and privately stated concern was the paucity of active troop formations within the interior military districts to contend with possible internal disorders.[64]

[62] Ibid., 49.
[63] Ibid., 49–50.
[64] Ibid., 58, 116–27.

The Jagged Transition

Although Alekseev and Palitsyn viewed themselves as traditionalists, their 1908-assessment marked an important evolution away from the Miliutin-Obruchev system. Their stress on depth of strategic deployments acknowledged both Russian weakness and the tsarist emphasis since 1906 on a worst-case threat scenario. Lack of French support during the Bosnian crisis only reinforced the overall sense of vulnerability and even pessimism. At the same time, the two officers' proposals for peacetime troop redeployments and military reform acknowledged "lessons learned" from the Russo-Japanese War. However, their emphasis on overwhelming threat and the attendant requirement for caution might be taken to extremes. The tendency to do just that accounted for much of the jagged transition between the Miliutin-Obruchev system and its Sukhomlinov-dominated successor.

An additional ingredient was discontinuity in personnel. Heads rolled during and after the Bosnian crisis, with General V. A. Sukhomlinov first replacing Palitsyn in late 1908 as chief of *GUGSh* and then replacing Rediger in early 1909 as war minister.[65] Sukhomlinov was an intelligent and experienced former military district (Kiev) chief who possessed the confidence of the tsar. A strong-willed cavalryman whose last active combat service dated to 1877–78, Sukhomlinov believed in loyalty to his sovereign and absolute unity of command. However, he was also party to a messy divorce that called his character into question, and his subsequent marriage to a free-spending divorcée left him vulnerable to gossip and in constant need of money. Since 1905, he and Grand Duke Nicholas Nikolaevich had differed over the nature and course of military reform, but Sukhomlinov grudgingly understood that royal blood trumped commoner status. Still, the war minister disliked competitors high and low, and he used his position whenever possible to eliminate or marginalize them, regardless of birthright or position. Between 1909 and 1911, even Prime Minister Stolypin recorded only two consultations with Sukhomlinov outside formal venues.[66] The prospects for collaboration became even worse after Kokovtsov replaced Stolypin in 1911.

[65] See the biographical sketch in D. N. Shilov, *Gosudarstvennye deiateli Rossiiskoi imperii* (St. Petersburg, 2001), 628–31.

[66] On Sukhomlinov's characteristics and style, see, William C. Fuller, Jr., *The Foe Within: Fantasies of Treason and the End of Imperial Russia* (Ithaca, 2006), 45–8; A. S. Lukomskii, "Ocherki iz moei zhizni," *Voprosy istorii*, no. 8 (August 2001), 92–3; and Polivanov, *Iz dnevnikov*, 39, 42–3, 50, 76.

Frequent access to the tsar and control over the War Ministry and the Supreme Certification Commission enabled Sukhomlinov to exert his authority over every corner of the ground force establishment, save those occupied by Grand Duke Nicholas Nikolaevich and the all-powerful military district chiefs. Sukhomlinov subjected real and potential enemies to retirement or reassignment. With tsarist approval, he de-fanged "the Young Turks" with remote troop postings and re-subordinated the service branch inspectors-general to the War Ministry. Sukhomlinov also moved quickly to bring *GUGSh* under the War Ministry, and by the end of 1910 he had reorganized "the brain of the army" to rationalize the internal division of functional responsibilities and to establish clear lines of subordination.[67] He wanted Alekseev as quartermaster general, or chief of plans and operations, but rotational assignment took Alekseev to Kiev as chief of staff of that military district. Instead, the war minister had to settle on Colonel (later General) Iu. N. Danilov, a polished military bureaucrat and a very intelligent peacetime regimental commander.[68] Danilov was neither a combat veteran nor a gifted strategist. But, during his tenure as quartermaster general between 1909 and 1914, his emphasis on orderly bureaucratic routine fit well with the war minister's sense of centralized authority. Between 1909 and 1914, the chiefs of the various functional sections within *GUGSh* increasingly came to resemble Danilov. Their attributes included a staff academy education or its equivalent and prior service either in central military administration or on military district staffs, especially Kiev. Personal acquaintance with Sukhomlinov was frequently a common characteristic. Except for Dobrorol'skii, who returned from regimental command to head the all-important troop mobilization section as a general after 1912, none had any recent direct combat experience.[69]

At the same time, Sukhomlinov employed the revolving door to preclude competition from his *GUGSh* chiefs. No fewer than four officers held this post during 1909–1914, and the last two, Ia. G. Zhilinskii and

[67] A. G. Kavtaradze, "Iz istorii russkogo general'nogo shtaba," *Voenno-istoricheskii zhurnal*, no. 7 (July 1972), 91–2; Sukhomlinov, *Vospominaniia*, 235; Lukomskii, "Ocherki iz moei zhizni," 93; Fuller, *The Foe Within*, 57; and Alekseeva-Borel', *Sorok let v riadakh russkoi imperatorskoi armii*, 267.

[68] V. A. Avdeev, "General Iu. N. Danilov i ego kniga," afterword in Iu. N. Danilov, *Na puti k krusheniiu*, ed. V. A. Avdeev (Moscow, 1992), 242–43; see also, *Spisok General'nogo shtaba*, 1 June 1914 ed. (Petrograd, 1914), 138.

[69] P. K. Kondzerovskii, *V stavke verkhovnogo 1914–1917* (Paris, 1967), 23–4; Lukomskii, "Ocherki iz moei zhizni," 103; and Skvortsov, Rushin, et al. (eds.), *General'nyi shtab Rossiiskoi armii*, 93–4.

N. N. Ianushkevich, were "chancery generals," who had made their reputations within the higher military bureaucracy in the imperial capital. Their skills and attributes suited the times, in which the larger institutional pattern was one of unchallenged bureaucratic authority within narrow ministerial and agency "stovepipes."[70]

The larger compartmentalized context also affected the course of military reform. The impact of the Bosnian crisis had left Prime Minister Stolypin with sufficient authority to impose a degree of unity on his often fractious cabinet, but the ministers of war and navy remained outside the collaborative fold, with important implications for the army. Throughout the second half of 1909, Stolypin repeatedly sought cooperation from the two ministries to harmonize their requests for extraordinary appropriations for key acquisitions. However, the navy would not relinquish its quest for dreadnoughts, and Sukhomlinov held the fortresses hostage to his own modernization program. As early as February 1909, the tsar had in principle agreed to scrap much of the fortress network, but Sukhomlinov was less than forthcoming about this arrangement as he fought for additional resources. Finally, in March 1910, Stolypin abandoned attempts at harmonization and simply mandated a budgetary compromise that imposed a roughly equal split (715 versus 698 million rubles) in extraordinary appropriations between the army and the navy.[71] The navy now had more elbow room to press forward with the construction of capital ships, including completion by 1914 of four dreadnoughts laid down in 1909 on the basis of annual operating funds. At the same time, Stolypin's Solomon-like decision precluded sufficient funds to modernize all the western fortresses, so Sukhomlinov now possessed a fiscal rationale for openly abandoning Warsaw (except for the citadel), Ivangorod, Novogeorgievsk, and Żegrze.[72]

Until additional supplements in 1913, the Stolypin-mandated compromise of 1910 set the budgetary terms for Sukhomlinov's military reform program. Building on initial appropriations that dated to 1908, the war minister was able to rearm the Russian army, complete modernization of its light artillery, procure machine guns (eight per regiment), and begin

[70] Lukomskii, "Ocherki iz moei zhizni," 94–5, 99; see also, A. F. Geiden, *Itogi russko-iaponskoi voiny 1904–05 gg.* (Petrograd, 1914), 81; and Golovin, *Plan voiny*, 76–8.

[71] Shatsillo, *Ot Portsmutskogo mira*, 149–59; see also, Gatrell, *Government, Industry and Rearmament in Russia*, 135–37.

[72] David Stevenson, *Armaments and the Coming of War: Europe, 1904–1914* (Oxford, 1996), 146–53.

the limited acquisition of aviation and motorized assets.[73] However, there were never enough funds to address all requirements, and two important objectives eluded the war minister. With sufficient means only for selectively rebuilding and rearming fortresses, he chose a "rear-to-front" option in the West that allocated scarce resources for Brest-Litovsk and Kovno, to which in 1912 he grudgingly added Novogeorgievsk. The remainder he would abandon, except for fortifications on the Narew crossing at Osowiec.[74] And, second, he lacked funds to complete the army's rearmament with heavy artillery, even though he understood that the Germans held a competitive advantage in firepower. By 1914, only one-half of the 37 Russian army corps was completely rearmed with modern heavy artillery systems.[75]

Other wounds were self-inflicted. The lessons from 1904–05 seemed to confirm prevailing wisdom, based in large part on perceptions from 1870–71, 1877–78, and 1904–05, that a European war would be short and decisive. Moreover, despite the importance of siege operations during 1904 at Port Arthur, the larger Manchurian context seemed to indicate the overriding importance of maneuver in modern war. The war minister and his writers of tactical doctrine took this lesson to heart, publishing a new set of tactical regulations in 1912 that duly emphasized the importance of maneuver in the field. However, the new manual ignored the complexities of perhaps the most difficult maneuver, the meeting engagement, in which opposing forces entered the initial phase of main battle from the march. And, thanks to Sukhomlinov's direct influence, tactical doctrine continued to emphasize direct frontal confrontation once forces were firmly in contact with the enemy.[76]

The lessons of 1904–05 also held important implications for munitions stockpiling and logistics. Authorities within the War Ministry attentively studied expenditure rates for artillery shells in Manchuria and

[73] V. A. Avdeev, "V. A. Sukhomlinov i voennye reformy 1905–1912 godov," in Rybachenok, Zakharova, and Ignat'ev (eds.), *Rossiia*, 276–77.

[74] Zaionchkovskii, *Podgotovka Rossii k imperialisticheskoi voine*, 146–49; Sukhomlinov, *Vospominaniia*, 173.

[75] E. Z. Barsukov, *Russkaia artilleriia v mirovuiu voinu*, 2 vols. (Moscow, 1938–1940), I, 18–20.

[76] Golovin, *Galitsiiskaia bitva*, 158–59, 338–39; see also, Bruce W. Menning, "The Offensive Revisited: Russian Preparation for Future War, 1906–1914," in David Schimmelpenninck van der Oye and Bruce W. Menning (eds.), *Reforming the Tsar's Army: Military Innovation in Imperial Russia from Peter the Great to the Revolution* (Cambridge, UK and New York, 2004), 216–19.

extrapolated figures from them in straight-line fashion to determine peacetime procurement requirements for a future maneuver war of short duration. The War Ministry generously added 250 shells per gun to projected requirements, settling on a predictable peacetime procurement figure of 1,000 per gun. It went unnoticed that under actual combat conditions some hotly engaged batteries had expended as many as several thousand rounds per day.[77] Reading the tea leaves for logistics requirements was similarly an inexact science. Nearly everyone from Sukhomlinov on down recognized the need to revamp the army's field logistics system. However, the maneuver war in Manchuria – episodic as it had been – yielded uncertain data on which to base extrapolations for the future. Therefore, limitations inherent in the cranky troop mobilization and railroad transit systems became the defining criteria for a reorganized logistics system. It would draw from a combination of forward magazines and centrally located depots linked to maneuver formations via transportation battalions. However, this solution held two significant shortcomings. First, the reserve troops who made up the bulk of the transportation battalions were located far back in the mobilization schedule. Thus, they could not make their presence felt until at least the third and fourth weeks of a future war. And second, during the interim, first-line maneuver units were tethered to their own organic support units. Their limited carrying capacities meant that logistics would become problematic any time the distance between supporting railheads and advancing maneuver formations exceeded two or three days' field march.[78]

Within the larger reform picture, the dictates for changes in army organization, structure, and peacetime deployments came from recommendations in the Alekseev-Palitsyn report. The tsar imposed only two conditions on Sukhomlinov's reform program: that reorganization must not increase the burden of military service; and that structural alterations must impose no new expenditures. In brief, Sukhomlinov's modifications simplified army organization, especially in the infantry, introduced the hidden cadre system for expandable reserve units into the structure of active formations, and prescribed a substantial peacetime redeployment of

[77] Barsukov, *Russkaia artilleriia v mirovuiu voinu*, I, 65–8; see also A. A. Manikovskii, *Boevoe snabzhenie russkoi armii*, 3 vols. (Moscow, 1920–1923), III, 7–10; and David R. Jones, "Imperial Russia's Forces at War," in Allan R. Millett and Williamson Murray (eds.), *Military Effectiveness*, 3 vols. (Winchester, MA, 1988), I, 262–63.

[78] N. N. Golovin, *Voennnye usiliia Rossii v mirovoi voine*, reprint ed. (Zhukovskii-Moscow, 2001), 272; see also, Ushakov, *Podgotovka*, 167–71; and Vasil'ev, *Transport*, 50–1.

Russian troops away from the western military frontier.[79] In accordance with changed peacetime dispositions, more than 200,000 troops were redeployed from the frontier military districts to the Moscow and Kazan military districts. These redeployments withdrew 91 battalions from the Warsaw district and 37 from the Vilnius district to spread peacetime troop dispositions more evenly across the Russian interior. The presence of additional active service formations in the interior served as antidote to rural and urban disturbances and eased the burden of mobilization and transit for possible war in the Far East, although threat assessments remained fixated on the West rather than the East. An added advantage of redeployment was economic: development of necessary infrastructure benefited not just Russian Poland, but also the interior provinces of the empire.[80] The great drawback was that redeployment to the interior complicated the calculus for mobilization and transit for war on the western frontier, a fact that did not escape the notice of the French General Staff.

Mobilization Schedule 1910 clearly demonstrated the twin effects of new troop deployments and pessimistic threat assessment in the West. The schedule was worst-case based and failed to observe economy of force. Moreover, the schedule's considerations provided for no sure transition from defensive to counteroffensive operations. Expanded requirements for troop mobilization and integration across more districts now drew forces from both the Caucasus and Siberia, but the deployment scheme frittered away new-found wealth by allocating two entire armies (three corps and six reserve divisions) to the defense of St. Petersburg and the border with Romania. Of the remaining available 24 active-army corps, 17 were deployed in four armies against Germany, leaving against Austria-Hungary only a single ungainly army of seven corps divided into two groupings. In a major departure from precedent, there were no front-level command instances, only an arrangement for general subordination of all armies to an overall supreme commander. In another marked departure from precedent, there was no explicit counteroffensive mission for any of the forward-deployed armies. The two groupings making up the single army concentrated south of the Pripet were to defend against Austria-Hungary, while the four armies north of the Pripet were

79 See the summary in Bruce W. Menning, "Mukden to Tannenberg: Defeat to Defeat, 1905–1914," in Frederick W. Kagan and Robin Higham (eds.), *The Military History of Tsarist Russia* (New York, 2002), 214–15.

80 RGVIA, f. 2000, op. 1, d. 6659, ll 41–10b.; Avdeev, "V. A. Sukhomlinov," 272–73; cf., Fuller, *Strategy and Power in Russia*, 426–32, 438–42.

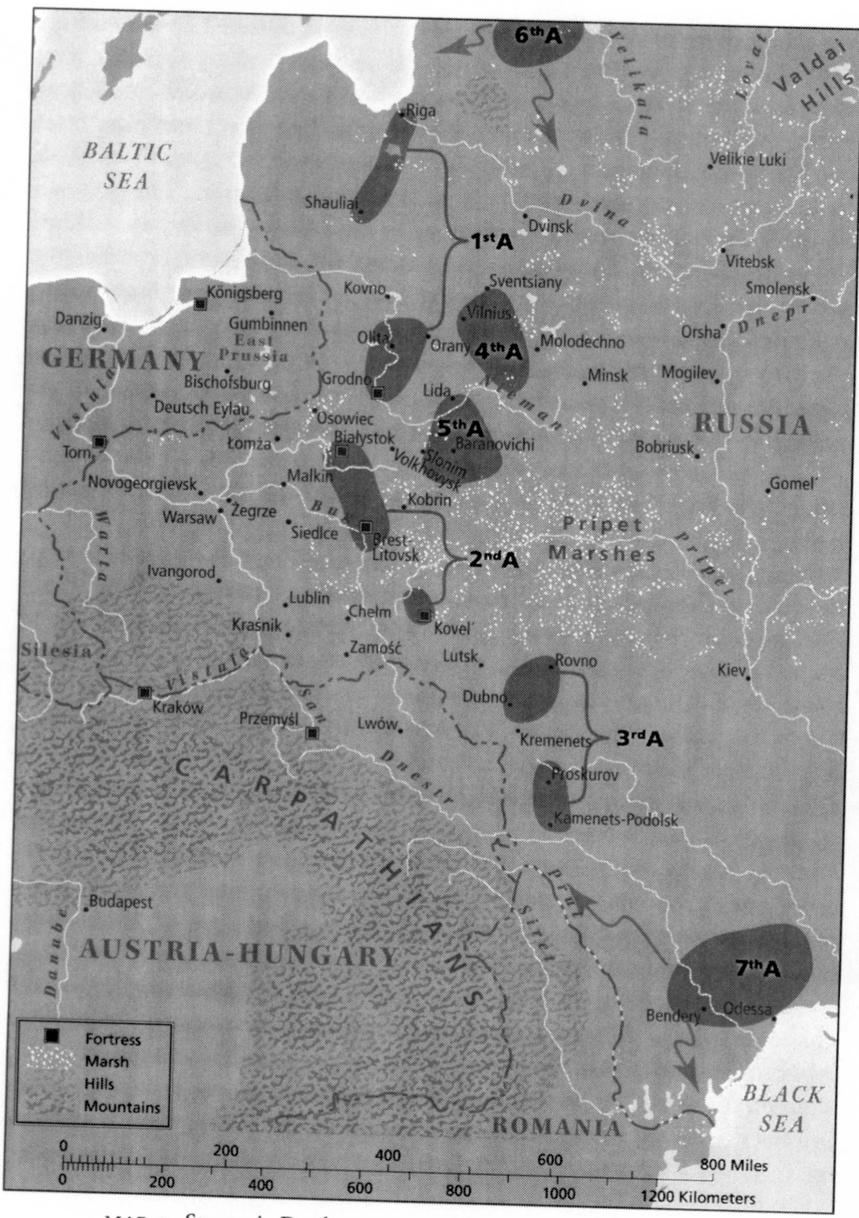

MAP 3. Strategic Deployments, Mobilization Schedule, 1910.

"to create as possible a favorable situation for transition with a combination of forces to a general offensive upon orders from the supreme commander."[81]

In military-strategic terms, the accent on defense and the primacy of the anti-German mission were less troubling than the location of anticipated strategic deployments and the difficulty of attaining mutual support and mass at key points of defense and possible counterattack. All army-level dispositions were on the shoulders of the Polish salient, eight to ten days' field march from the middle Vistula. The four armies north of the Pripet were arrayed two each in two strategic echelons, leaving insufficient deployment and maneuver space to bring all combat power to bear (except sequentially) against German forces in East Prussia. Meanwhile, planners within both the Warsaw and Kiev military districts understood that the lone Russian army (seven corps) south of the Pripet might expect an onslaught from as many as 13 Austro-Hungarian corps. The corresponding two Russian groupings between Rovno-Dubno and Proskurov ceded too much space and invited defeat in detail. Worse, planners worried about sabotage and even armed insurrection among elements of the Polish population that had grown weary of Russian occupation. Still worse, planners were fearful that Austro-Hungarian troops advancing against light resistance inside the Polish salient would turn north to attack the rear areas of Russian forces deployed against East Prussia.[82] Even as Mobilization Schedule 1910 took effect, the frontier district staffs and *GUGSh* were at odds over how to palliate serious shortcomings in the strategic concept.

Beyond the strategic calculus, Mobilization Schedule 1910 held a small host of political-military repercussions. Despite official circumspection, word leaked out that frontier fortresses were being demolished or abandoned. Consternation boiled over from the Duma into the Council of Ministers and into the upper reaches of the officer corps, including the headquarters of Grand Duke Nicholas Nikolaevich. The majority of senior officers had matured under the Miliutin-Obruchev system, and there was scant sentiment for leaving ten provinces of Russian Poland only lightly defended in the event of enemy invasion.

There was also the problem of the alliance with France. The abandonment of the fortresses, together with the peacetime redeployment of

[81] RGVIA, f. 2000, op. 1, d. 1790, l. 293; the entire set of considerations appears on ll. 293–297.

[82] For an extensive military critique of Mobilization Schedule 1910, see, ibid., ll. 49–59ob.

active units from the western military frontier into the Russian interior, held worrisome implications that found their way into French newspapers and that deeply disturbed the French General Staff. Pressures soon built, forcing War Minister Sukhomlinov to dispatch personal emissaries to Paris with reassurances of Russian commitment to the alliance.[83]

At the same time, intelligence-based assessments figured prominently in building planning sentiment against the new mobilization schedule. Between 1909 and 1912, thanks to espionage penetration of the Austro-Hungarian high command, *GUGSh* gradually acquired high-grade intelligence on Fall R, Vienna's scheme for deployments and operations against Russia in the event of a European war. The materials were sufficiently detailed to reveal the locations of corps-level formations and their initial axes of advance into Russian territory.[84] Possession of such information underscored at least two significant issues for war planners both in *GUGSh* and within the staffs of the military districts: the mismatches between Russian dispositions and those of the Austro-Hungarian adversary; and the opportunity costs inherent in the failure to profit from an intelligence coup.

As planners pondered these questions, along with knowledge there was no corresponding hard intelligence on German plans and dispositions, events on the international scene called into question various postulations and assumptions from earlier pessimistic threat assessments. After the Bosnian crisis, Balkan tensions subsided for a time, as Russia and Austria-Hungary normalized relations. In late 1910, the tsar and Foreign Minister Sazonov papered over differences with Berlin and acquiesced to German construction of the Berlin-Baghdad Railroad in exchange for recognition of a Russian sphere of influence in northern Persia. More significantly, the Second Moroccan Crisis erupted in mid-1911, fanning the winds of war between France and Germany until a compromise exchange

[83] V. A. Apushkin, *General ot porazhenii V. A. Sukhomlinov* (Leningrad, 1925), 38; Pertti Luntinen, *French Information on the Russian War Plans 1880–1914* (Helsinki, 1984), 122–28; and Fuller, *Strategy and Power in Russia*, 432–33.

[84] For a copy of Fall R, see, RGVIA, f. 2000, op. 1, d. 1774, ll. 100–110ob. The source is generally believed to have been Colonel Alfred Redl, former chief of counterintelligence for the Austro-Hungarian General Staff. However, Mikhail Alekseev in *Voennaia razvedka Rossii*, 3 vols in 4 bks. (Moscow, 1998–2001), II, 191–97 makes a strong case for the fact that historians still do not know the exact identity of Agent 25. See also, Alex Marshall, "Russian Military Intelligence, 1906–1917: The Untold Story behind Tsarist Russia in the First World War," *War & History*, XI, no. 4 (November 2004), 397–98; and John R. Schindler, "Redl – Spy of the Century?" *International Journal of Intelligence and Counterintelligence*, XVIII, no. 3 (Fall 2005), 486–88, 490–95.

of colonial concessions resolved the dispute. Interestingly, Russia, perhaps as riposte for lack of assistance from its ally during the Bosnian crisis, offered France only ambiguous assertions of support. To Germany's surprise, Great Britain more than compensated by throwing its weight behind France. In September 1911, as the Moroccan crisis entered its final phase, Italy invaded Libya, igniting the Tripolitan War against Turkey. Concerned with maintaining the status quo in the Balkans, Austria-Hungary acted to prevent Italian attacks against the Turkish littoral in the Adriatic and Aegean. Meanwhile, Russia sought its own version of stability by fostering an alliance of Balkan states, including Turkey, to serve as a bulwark against Austro-Hungarian incursion. In 1911, news that Turkey might purchase several dreadnoughts abroad motivated the Council of Ministers and the Duma to authorize additional scarce funds for the construction of three Russian dreadnoughts on the Black Sea.[85]

A resurgent Turkish threat merely complicated matters for Russian naval officers who wrestled with their own version of contingency planning for a European war. With Russia's first four dreadnoughts not due for completion until late 1914, and with a patchwork quilt of pre-dreadnoughts and secondary vessels of various vintages, the Naval General Staff harbored few illusions about the capabilities of the Baltic and Black Sea fleets. The emphasis was on the defensive in planning, and both the army and navy agreed – for once – that the near-term tasks for the Baltic Fleet centered on the defense of St. Petersburg and the nearby coast of the Finnish Gulf from bombardment and enemy landings. In any confrontation at sea against Germany, the Russians figured to be heavily outnumbered and outgunned. Therefore, the mission of the Baltic Fleet was to buy time, as much as one or two weeks, so that the army might mobilize sufficient ground forces to defend against German and possible Swedish incursions. Utilizing the limited assets at its disposal, the Naval General Staff by 1912 settled on a combination of mine barriers, pre-dreadnoughts, cruisers, submarines, and torpedo boats to bear the main burden of Baltic defense.[86] With some adjustments during the spring of 1914, these preparations governed Russian naval operations in the Baltic during the early months of World War I.

[85] See the overview in Stevenson, *Armaments and the Coming of War in Europe*, 180–95, and 225–43.

[86] Rossiiskii gosudarstvennyi arkhiv voenno-morskogo flota [hereinafter RGAVMF], f. 418 (*Morskoi General'nyi shtab*), op. 2, d. 221, ll. 8–19; and d. 215, ll. 37–59ob.; see also, Petrov, *Podgotovka*, 206–09; and Podsoblyaev, "The Russian Naval General Staff," 57–9.

Preparation for war on the Black Sea presented a different set of dilemmas. Missions for the Black Sea Fleet oscillated between maintaining command of the sea and forming the backbone for a resurrected joint army-navy expedition against the Bosporus. These two missions were not mutually exclusive, but wherever the primary emphasis fell, the chief recurring problem was uncertainty of threat. Like its Baltic counterpart, the Black Sea Fleet could not count upon dreadnought-style augmentation until well into the second decade of the twentieth century. Until 1912, when it became ever more likely that the Ottomans would purchase two dreadnoughts abroad, the Russians held superiority in any potential one-on-one confrontation against the Turks. However, possible contention at the Straits would probably draw in naval combatants from one or more of the European powers, in which case the Russians surrendered dominance. It was for this reason that various plans for a Bosporus expedition remained conservative in nature. The aim was to deny either the Turks or an outside power the ability to wrest mastery of the Black Sea from Russia. Therefore, the Russian concept was to sail to the outlet of the Bosporus to the Black Sea, lay down a mine barrier, and then land an army corps to secure the mine barrier from the landward side and to cover the barrier with heavy artillery. However, there always seemed to be too many imponderables for navy and army planners to work out the Bosporus expedition as a viable course of action. The situation worsened in 1913, when the Russians learned that Turkey was on the verge of purchasing two additional dreadnoughts, and that Germany was intent on transferring two modern cruisers to the Ottoman navy at the outset of any possible Russo-Turkish hostilities.[87]

In light of these and other complications, planning for a Bosporus expedition became a paperwork exercise, while planning for sea mastery assumed a conservative, even passive nature. By 1914, the governing concept was to maintain from Sevastopol only command of the northern reaches of the Black Sea. Mine barriers would be laid down in select locations, while the Black Sea Fleet remained concentrated at its base. A combination of means, including mines, cruisers, torpedo boats, and submarines, would wear down opposing forces to parity with the Russians, at which point the Russian fleet would sally to wrest command of the entire Black Sea. Above all, in the event of a European war, the Black Sea Fleet was to refrain from provocative actions that might draw Turkey

[87] RGAVMF, f. 418, op. 2, d. 254, ll. 1–6; and Petrov, *Podgotovka*, 234–36.

into the ranks of Russia's opponents.[88] This constraint did not vary much from the intent behind instructions of early 1904 that had so hampered the Russian Pacific Squadron in its preparations for a possible war against Japan.

The Revolt of the Generals

The evolving threat calculus in all its dimensions fueled dissatisfaction in several quarters with Mobilization Schedule 1910. Criticisms of its underlying strategic considerations drew strength from the staffs of the Kiev and Warsaw military districts and support from Grand Duke Nicholas Nikolaevich. The intellectual spark for revision came from the respective chiefs of staff of the two districts, General Alekseev and General N. A. Kliuev. Although the trend to corporate-style exclusiveness within *GUGSh* tended to insulate the central planning apparatus from the periphery, the center could not totally ignore the power of the military district chiefs, who were subordinate only to the throne. Their staffs, in turn, might benefit from their chiefs' stature to emphasize collaborative planning in a way that challenged the rote imposition of Mobilization Schedule 1910. The result was a generals' revolt that culminated during early 1912 with substantial changes to the Russian strategic concept for engagement in a general European war.

Revolt found its remote origins in dissatisfaction with orchestrated proceedings following the cancellation in December 1910 of a strategic war game that would have put the considerations that underlay Mobilization Schedule 1910 to the test. At the last minute, *GUGSh* replaced the game with a series of conferences during which Alekseev's critical comments about the new schedule and its underlying strategic concept were ignored.[89] He subsequently spent the next year poring over maps and intelligence summaries to develop a new strategy, which he labeled the "General Plan of Actions." In two succinct pages, Alekseev deftly outlined a powerful rationale for changed Russian strategic deployments and courses of action. Although intelligence on Fall R loomed not far in the background, he built his case for changed priorities on the altered

[88] V. A. Zolotarev, "Deistviia russkogo flota v Pervoi mirovoi voine," *Voenno-istoricheskii zhurnal*, no. 1 (January 2002), 56; Petrov, *Pogotovka*, 241–47.

[89] Alekseeva-Borel', *Sorok let v riadakh russkoi imperatorskoi armii*, 286–88; and Polivanov, *Iz dnevnikov*, 101.

MAP 4. Alekseev's General Plan of Actions, 1912.

geopolitical environment. In Alekseev's view, Italy's commitment during 1911 to a colonial war in Africa, together with deep-seated animosities against the Austrians, clearly rendered Rome an unreliable partner in the Triple Alliance. It followed that Italy would not likely commit military forces against France in a future European war.

At the same time, Alekseev perceived that naval and commercial rivalries between Germany and England were inexorably driving the latter into the Franco-Russian camp. In the event of war, Germany could scarcely leave only a ground-force screen in the West against France and England. More importantly, Alekseev held the Germans understood that war in the East against Russia could scarcely be decided by a single stroke, "however low the Germans rate our higher command personnel." Alekseev concluded that "for Germany it was both dangerous and disadvantageous to be drawn into a protracted struggle on the eastern front." For the Germans it would be better to observe economy of force in the East by drawing superior Russian forces into a protracted fight for East Prussia, a sub-theater that at best lay only on the flank of primary operational lines in the direction of Berlin.

For Alekseev, these important considerations called for a timely review of "the general idea of war." His intent was to alter that idea so that during the initial period of a European war Russia would direct its principal forces against Austria-Hungary. Against Germany he would leave a mere six corps concentrated between Grodno and Białystok. In the end, Alekseev held that Austria-Hungary indisputably remained Russia's primary enemy, and, in view of force correlations, the more dangerous. Besides, success against Austria-Hungary promised palpable short-term results. They included quieting the restive population in Russian Poland, subverting the subject Slavic populations of Austria-Hungary, and distancing the war from the Russian frontier. Thus, it was against the Habsburgs in his words that "our forces ought to be directed, decisively and without hesitation."[90] In January 1912, Alekseev discussed these views with the grand duke. The intent was to propose adoption of Alekseev's plan the following month during a conference of the military district chiefs of staff in Moscow.

Alekseev was not without competition. In January 1912, Quartermaster General Danilov outlined an alternative strategic concept that emphasized the same changes in the international situation noted by Alekseev.

[90] The plan is at RGVIA, f. 2000, op. 1, d. 2309, ll. 112–13; see also, Zaionchkovskii, *Podgotovka Rossii k imperialisticheskoi voine*, 236–37.

However, unlike Alekseev, Danilov repeated the same Germany-first emphasis from earlier considerations, producing two variations on the same theme for operations against Germany: offensive and defensive. In addition, he spelled out in greater detail than Alekseev the prevailing notions that underlay the impulse for change: Mobilization Schedule 1910's emphatic emphasis on the defensive and the altered international climate, especially England's growing affinity for the Franco-Russian alliance after the Second Moroccan Crisis. Danilov cited an imminent Anglo-French naval convention that would split the two nations' maritime areas of responsibility, thereby freeing the preponderance of England's battleship force for concentration in the North Sea against Germany. There was also great likelihood that in any large-scale European conflict the British would commit a 100,000-man expeditionary force to the Continent against Germany. Together with the advent of a liberal government in Sweden, these developments held at least two significant strategic implications. First, Germany would more likely than ever turn westward at the outset of future conflict, and second, there was decreased probability that either Germany or Sweden would launch large-scale operations against either the vulnerable Baltic coast or St. Petersburg. In light of these and other assessments, Danilov would initially allocate three full field armies against Germany, with the provision to reinforce with two additional armies early in the second month of war. He would commit only seven corps against Austria-Hungary. He made only passing mention to the possibility of striking first against Austria-Hungary.[91] Thus, he failed to address two serious deficiencies in Mobilization Schedule 1910: the lack of maneuver space in the North to bring total force to bear against German dispositions in East Prussia, and the negative force correlations against Austria-Hungary (at least two-to-one against Russia). Lopsided odds in the South meant Russian troops deployed in the North against East Prussia would be vulnerable in their rear to a concerted Austro-Hungarian offensive into Russian Poland. These considerations seriously weakened Danilov's case.

Two additional elements of Danilov's proposal merit attention in light of what actually occurred during initial Russian offensive operations in August 1914. First, in contrast with Alekseev, Danilov emphasized the necessity to attain speed in strategic deployments and initial operations. If Germany turned first against France, Danilov expected the French to hold out no longer than two months, after which Berlin would direct

[91] RGVIA, f. 2000, op. 1, d. 1813, ll. 4–4ob., 7, 9, 15–16.

its undivided fury eastward. This assumption Danilov possibly derived from Russian intelligence on exchanges between the chiefs of the German and Austro-Hungarian general staffs. Second, to attain speed, Danilov advocated engagement in initial offensive operations before completion of mobilization and concentration. Alekseev had earlier contemplated the same hazardous enterprise, but later events would demonstrate that he was far more circumspect in application.[92]

Finally, alliance considerations occupied an important place in the thinking of both officers, although only implicitly for Alekseev. Danilov was explicitly committed to direct support of France by means of a full-scale attack into East Prussia. However, limited railroad infrastructure and restrictive deployment and maneuver space levied several implied tasks: first, there was a requirement to press no fewer than two armies into East Prussia, one from the east across the Niemen and the other across the Narew from the north face of the Polish salient; and second, there was a requirement after initial operations for the regrouping of these and follow-on Russian assets to press sequential operations across the lower or middle Vistula into the heartland of Germany. It was the necessity to surmount these obstacles that caused Alekseev to view initial operations against East Prussia as only tangential to the primary anti-German axis of advance across the middle Vistula, the more so since such operations would be vulnerable from the rear to advancing Austro-Hungarian forces attacking from the south face of the Polish salient.

Alekseev, therefore, would opt for his own version of sequential operations, but with a far different immediate objective, Austria-Hungary. The Russians possessed the plan for Austro-Hungarian deployments, and the multi-ethnic nature of the Habsburg army rendered it more vulnerable than the Germans to rapid defeat and disintegration. The only logical course of action for Alekseev was to launch an immediate knock-out blow against Austro-Hungarian forces in Galicia, and then to regroup either for an advance through the Carpathians or for an advance across the middle Vistula into Germany and on to Berlin. Although Alekseev

[92] Ibid., ll. 90b., 130b. Whatever the role and fate of Colonel Redl, the Russians retained what appeared to be a high-ranking intelligence source inside the Austro-Hungarian Ministry of War, and, until the early months of 1914, reports from this agent went through the Kiev military district to *GUGSh* and the tsar. However, the Russian spymaster N. S. Batiushin, in *Tainaia razvedka i bor'ba c nei*, reprint ed. (Moscow, 2002), 79–80, discounted the veracity of the reports. For examples of them, see, RGVIA, f. 2000, op. 1, d. 1774, ll. 250–51; and "K voprosu o podgotovke mirovoi voiny," *Krasnyi arkhiv*, no. 64 (1934), 90–3.

would also come to the assistance of France, his route to the objective was more circuitous.[93] It did not hurt that the humiliation of 1908–09 had added still another score to settle against Austria-Hungary.

Danilov's emphasis on the anti-German option demonstrated both fear of German capabilities and an increasingly narrow interpretation of alliance obligations. During 1911, Russia had been less than forthcoming in its support for France during the Second Moroccan Crisis, and in August, the allied chiefs of staff met at Krasnoe Selo, even as the crisis was in full swing. In the event of war with Germany, General Auguste Dubail assured General Zhilinskii that on the twelfth day of mobilization the French army would be ready for offensive operations "with the assistance of the English army on its left flank." Zhilinskii lamely replied that the Russian army was still undergoing reform. Therefore, he promised only that by the fifteenth day Russia would deploy sufficient troops to tie down five or six German corps in the East. A year later in Paris, Zhilinskii promised the new French chief, General Joseph Joffre, that Russia would deploy 800,000 troops on the German frontier with the intent to engage in offensive operations "after the fifteenth day of mobilization."[94] Reality would later reveal a substantial gap between these figures and Russian capabilities, but Zhilinskii's promise was grounded in substantial alterations made during early 1912 to the considerations inherent in Russian strategic deployments.

The Evolution of the Sukhomlinov System

When the chiefs of staff of the military districts assembled in Moscow during February 1912 to review the strategic and logistical concepts that underlay Mobilization Schedule 1910, Alekseev successfully proposed his "General Plan of Actions" to replace the earlier and overwhelmingly defensive considerations. However, his victory was not unqualified. When the war minister and the district chiefs subsequently convened to sanction the Alekseev-inspired departure, their approval favored Alekseev, but left room for Danilov. As endorsed by the war minister and forwarded to the tsar, the collective wisdom of the assembled officers was to direct Russia's primary effort against Austria-Hungary, but not to preclude simultaneous offensive operations into East Prussia. Thus,

[93] See the commentary in Golovin, *Galitsiiskaia bitva*, 10–14.

[94] Zaionchkovskii, *Podgotovka Rossii k mirovoi voine v mezhdunarodnom otnoshenii*, 235–36, 279–80.

Russia might serve its traditional anti-Austro-Hungarian interests while simultaneously addressing a narrowly-interpreted version of alliance obligations.[95] Sukhomlinov's decision was to compromise, and in compromise was born a rationale for simultaneous major offensive operations across two diverging and non-mutually supporting strategic axes. This compromise the military commentator General N. N. Golovin would later label "the worst decision of all."[96]

A second major mistake – implicit in compromise – was the failure to allocate decisive mass to either axis. This failure took root gradually during the remainder of 1912 and over 1913 under the guise of refinements and changes to Alekseev's original concept. Subsequently, an obsession with speed, apparent during rehearsal and actual application in 1914, would magnify the failure to plan for attainment of mass against either potential adversary. At the same time, logistical shortcomings imperiled the capacity to sustain and support mass, even in an attenuated form.

During 1912, Alekseev and his counterpart in Warsaw, General Kliuev, bore primary responsibility for refining the new strategic concept in all its dimensions. However, several considerations hamstrung their collaborative approach to war planning. First, planning occurred inside the framework of the standing mobilization schedule from 1910, which came to be known in short-hand parlance as Schedule No. 19. A new schedule, No. 20, was not due until the spring of 1914. Second, guidance from above limited – even diminished – the number of corps-level assets available for operations on the Southwest Front against Austria-Hungary. Although collaboration was short-lived, it was both lateral and vertical, and Danilov from his vantage at *GUGSh* was not above using his authority and influence to restrict Alekseev's freedom of planning action. And, third, all planners high and low had to contend with the tsarist emphasis on worst-case planning.

As a result, the strategic concept that flowed from planning priorities during 1912 would have two variants, "A" for a major offensive effort against Austria-Hungary, and "G" for a contrasting major effort with a defensive emphasis against Germany, should that country's armed forces initially turn east. Planners paid only lip service to the latter proposition, with the result that "Plan G" was never as fully developed as its

[95] See the overview in Bruce W. Menning, "Pieces of the Puzzle: The Role of Iu. N. Danilov and M. V. Alekseev in Russian War Planning before 1914," *The International Historical Review*, XXV, no. 4 (December 2003), 794–96.

[96] N. N. Golovine, *The Russian Campaign of 1914. The Beginning of the War and Operations in East Prussia* (Ft. Leavenworth, KS, 1933), 67.

anti-Austro-Hungarian counterpart.[97] In August 1914, Russia would go to war on the basis of "Plan A" as approved by the tsar on 1 May 1912, but subsequently altered during 1913 and on the fly in August 1914 (N. S.).

According to "Plan A," the initial mission for all Russian forces during concentration was "transition to the offensive against the armed forces of Germany and Austria-Hungary, with the objective of taking the war into their territory." For this purpose, "A" would allocate against Germany two armies (the First and Second) with 481 infantry battalions and 316 cavalry squadrons, the overall mission of which was "the defeat of residual German troops in East Prussia and its conquest, with the objective of creating an advantageous departure point for further actions." Against Austria-Hungary, "A" would allocate three armies (the Third, Fourth, and Fifth) with 744 battalions and 594 squadrons, the overall mission of which was "the defeat of the Austro-Hungarian armies, while preventing the movement of significant enemy forces south of the Dniester or west to Kraków." Two additional armies, the Sixth and Seventh, were allocated respectively to the defense of St. Petersburg and the border with Romania.[98] Under the overall scheme, the peculiarities of the Russian troop mobilization system meant that only 350,000 (of the alliance-mandated 800,000) Russian troops would be available by M+15 (mobilization day plus fifteen) for deployment against Germany.[99] Meanwhile, the three armies allocated to Southwest Front would leave Alekseev with only a razor-thin margin of superiority over the Austro-Hungarian forces that intelligence reports indicated might be deployed against the Russians.

Danilov would subsequently make the margin even thinner in pursuit of his own version of planning "rollback." In early 1913, he successfully importuned *GUGSh* chief Zhilinskii to curtail collaborative planning, thereby restoring the center's ascendancy.[100] However, to some extent, the damage to the periphery had already been done, because during 1912 Alekseev had rotated from staff to corps command at Smolensk, and Kliuev was on the verge of a similar reassignment to the Caucasus. Their departures obscured near-unanimous unease within the military district staffs over logistics and military-administrative arrangements, as army rear areas were shifted to accommodate revised concentrations and

97 RGVIA, f. 2000, f. 1759, op. 3, d. 610, ll. 1–3ob; and f. 2000, op. 1, d. 207, ll. 65–6ob.
98 RGVIA, f. 2000, op. 1, d. 1824, ll. 1–3.
99 Zaionchkovskii, *Podgotovka Rossii k imperialisticheskoi voine*, 321–22.
100 RGVIA, f. 2000, op. 1, d. 1819, ll. 49ob.-52.

deployments. Similarly, there was near-unanimous unease over inadequate stocks of small arms and artillery munitions, neither of which fully matched the requirements from earlier-established norms.[101] Kliuev in particular worried about extended army frontages on the potential Northwest Front that might imperil lateral linkages and command and control.[102] The worst blow against Alekseev's concept, however, came on 25 September 1913, when *GUGSh* issued a fresh set of imperially sanctioned "considerations" that became guidance for the overarching strategic concept until the advent of Mobilization Schedule No. 20 (now postponed until 1915). These considerations shifted an additional corps from Southwest Front to Northwest Front and prohibited any further shifting of corps-level formations from one front to the other. In compensation, Southwest Front nominally received an additional army, the Eighth, but it comprised only the Proskurov Grouping from the original iteration of Plan A. Meanwhile, the new considerations tied formations from the Warsaw military district more firmly to Northwest Front and emphasized speed at the expense of mass during initial operations.[103] The cumulative effect was to weaken Southwest Front and to distribute its corps-level deployments more evenly across its entire breadth. With less mass for Alekseev's concept in key sectors against Austria-Hungary, Danilov lamely emphasized the importance of Russian bravery in the attainment of victory over a multi-ethnic army susceptible to defeatism and disintegration.[104]

A corresponding emphasis on speed to compensate for the absence of mass found its most complete prewar expression during the strategic war game held in Kiev during April 1914. With the exception of several army-level commanders, the war game's participants counted nearly all the key *dramatis personae* who would figure in the actual onset of operations during August 1914. The considerations of 1 May 1912 as amended on 25 September 1913 established the strategic framework for troop deployments.[105] At notional M+9, the game forced Northwest Front to parry a German spoiling attack from East Prussia and then

[101] Ibid., d. 1803, ll. 315–17; and op. 3, d. 207, ll. 14–14ob, and 65–65ob.

[102] 10-i otdel General'nogo shtaba RKKA, *Vostochno-Prusskaia operatsiia. Sbornik dokumentov* [hereinafter *Vostochno-Prusskaia operatsiia*] (Moscow, 1939), 38–9.

[103] RGVIA, f. 1759, op. 3, d. 658, ll. 4–5, 9ob.-11ob.; and Ia. K. Tsikhovich (comp.), *Strategicheskii ocherk voiny 1914–1918 g.g.* (Moscow, 1922), 16–20.

[104] See the commentary in A. S. Beloi, *Galitsiiskaia bitva* (Moscow, 1929), 23–4.

[105] RGVIA, f. 1759, op. 3, d. 1059, ll. 91–2, 104.

to engage at M+12 in counteroffensive operations before the completion of mobilization and concentration. At M+14, there was a delay in these operations to permit First and Second armies to advance in concert against German dispositions forward of the Masurian Lakes. Southwest Front, meanwhile, was limited to concentrating its armies in expectation of transition to the offensive at M+20-M+21. The game's third turn brought news that a British expeditionary force had landed on the Continent, an event the Russians deemed sufficiently significant to trigger a German withdrawal to the Angerapp and the dispatch of three corps from East Prussia to the West. Southwest Front received orders to advance the date of its offensive by one day, and the last turn of the game found both Russian fronts in headlong pursuit of offensive operations before completion of concentration and strategic deployments. Participants complained about the emphasis on speed, but their concerns fell on deaf ears. Lesser commentary noted the difficulty of pursuing coordinated operations on either side of the Masurian Lakes.[106]

There was an important lacuna in the strategic war game. Rail transit and logistics were simply deleted in order "not to complicate the play." Thus, problems in supply and support that planners had already identified in 1912 received no additional attention, and the war minister failed to address them in his perfunctory commentary on lessons drawn from the game. Only individual participants noted apparent but untested deficiencies in logistics in their various after-action reports. Similarly, the game did not address failures to attain combat mass in selected sectors on Southwest Front, since the proceedings drew to a close before that front witnessed more than the last two turns of play.[107]

The game, together with the flow of planning during 1912–1913, revealed in broad outline what some observers later came to call the Sukhomlinov System. Unlike the Miliutin-Obruchev System, there was little flexibility to buy time and less leeway for the calculated pursuit of self-interests. A mixture of fear and opportunism (the 60-day window), combined with altruism (narrowly conceived alliance commitments), defined the requirement to initiate offensive operations against Germany at M+15. Self-preservation (the retention of Russian Poland) and a different version of opportunism (intelligence on enemy dispositions

[106] The game is summarized in A. N. Suvorov, "Voennaia igra starshikh voiskovykh nachal'nikov v aprele 1914 goda," *Voenno-istoricheskii sbornik, vyp.* 1 (1919), 15–19, 21–2; see also, Bruce W. Menning, *Bayonets before Bullets: The Imperial Russian Army, 1861–1914* (Bloomington, 1992), 251–54.

[107] Suvorov, "Voennaia igra," 22–4; see also, V. A. Melikov, *Strategicheskoe ravertyvanie* (Moscow, 1939), 275–76.

and perceived enemy weaknesses) defined the same kind of requirement – with somewhat more latitude – against Austria-Hungary. Without the forward fortresses, the Russians might press initial troop concentrations only midway into the Polish salient, and then only partially and at the expense of risk. In light of German and Austro-Hungarian advantages in speed of mobilization and concentration, the Russians early-on required sufficient mass forward to cover concentration and to blunt enemy probes and spoiling attacks. This mass the Russians might draw from the three frontier military districts, but Northwest Front held priority of assignment for troops from both the Vilnius and Warsaw districts. Southwest Front drew only residual troops from the Warsaw district, while its two vulnerable right-hand armies came from the more distant Moscow and Kazan military districts. Only Southwest Front's two left-hand armies were based on the nearby Kiev military district. Under these circumstances, prudence might dictate deferral of offensive operations until the complete buildup of combat power, but the Russians chose to counter speed with speed. That is, they chose to accept serious additional risk by engagement in full-blown offensive operations against both adversaries before the completion of mobilization and concentration.

Other departures from Miliutin-Obruchev were implicit in the locales for strategic troop deployments. During 1912, Alekseev had wanted to concentrate sufficient combat power on Southwest Front between Lublin and Chełm to smash and roll over Austro-Hungarian armies from the Northwest in a gigantic reincarnation of the classical Greek confrontation between Thebes and Sparta at Leuctra in 371 B.C. During 1913, Danilov would thin out and laterally extend Alekseev's concentrations with the implied intent to execute a modern version of Hannibal's victory of double envelopment over Rome at Cannae in 216 B.C., possibly north of Lwów (Lemberg). At the same time, Danilov would strengthen the two armies of Northwest Front to pursue the same kind of quest for Cannae against German forces on either side of the Masurian Lakes.[108] More than Alekseev, Danilov relied on speed to compensate for mass. Also, more than Alekseev, Danilov overlooked logistics, probably because initial objective depths were not great. To both officers' credit, there was the expectation for sequential operations, either into the Austro-Hungarian heartland or against Berlin. However, these operations would follow decisive initial confrontation, after which there would occur a regrouping of combat assets and a reorganization and displacement forward of logistics

[108] See the discussion in Golovin, *Plan voiny*, 89–93, 96–99; and in Golovin, *Galitsiiskaia bitva*, 10–12, 30–34.

assets. There was little or no sense that these processes might occur while still in contact with regrouping enemy forces.

Mobilization and Improvization

Once declared on 17 (30 July), full Russian troop mobilization proceeded without a major technical hitch, summoning some 3.4 million reservists to the colors over roughly 41 days in accordance with various modifications to Mobilization Schedule 1910, termed at the time Mobilization Schedule 19A.[109] However, difficulties and uncertainties appeared immediately.

Several urban centers witnessed draft riots, and shortages of officers and equipment plagued second-line reserve units.[110] Worse, as transit from assembly to concentration wound its lengthy course, the prevailing emphasis on speed began to exact its toll. Combat formations built around first-line (active) army units with direct reserve augmentation occupied first place in the deployment choreography, with the result that logistical assets and pure reserve infantry units lagged far behind.[111] Still worse, *Stavka*, or the Headquarters of the Supreme Commander, introduced last-minute changes into the strategic deployment scheme even as troops were still in transit to concentration. Thus, *Stavka* deprived the First Army on Northwest Front of I Corps and replaced the Guards Corps with XX Corps from Southwest Front. The object was to press the Guards and I Corps farther into the Polish salient, so they might constitute the nucleus of a new army, the Ninth, in the vicinity of Warsaw. *Stavka* was getting ahead of itself – the new army constituted the initial phase of regrouping for operations across the middle Vistula, even though operations against East Prussia had not yet commenced. Meanwhile, the left flank of Northwest Front's Second Army was shifted farther to the West, thus dangerously extending its frontage and aggravating an already serious command and control situation for an army whose command and staff personnel had never previously worked together.[112]

109 Dobrorol'skii, *Mobilizatsiia russkoi armii v 1914 godu*, 100; RGVIA, f. 1 (*Kantseliariia Voennogo ministerstva*), op. 2, d. 115, l. 23ob.

110 Josh Sanborn, "The Mobilization of 1914 and the Question of the Russian Nation: A Reexamination," *Slavic Review*, LIX, no. 2 (Summer 2000), 271–73.

111 Dobrorol'skii, *Mobilizatsiia russkoi armii v 1914 godu*, 104–06; S. M. Belitskii, *Strategicheskie reservy* (Moscow-Leningrad, 1930), 63–4.

112 [V. E. Borisov] *Kratkii strategicheskii ocherk voiny 1914–1918 gg.*, 2 vols. (Moscow, 1918–1919), I, 18, 20.

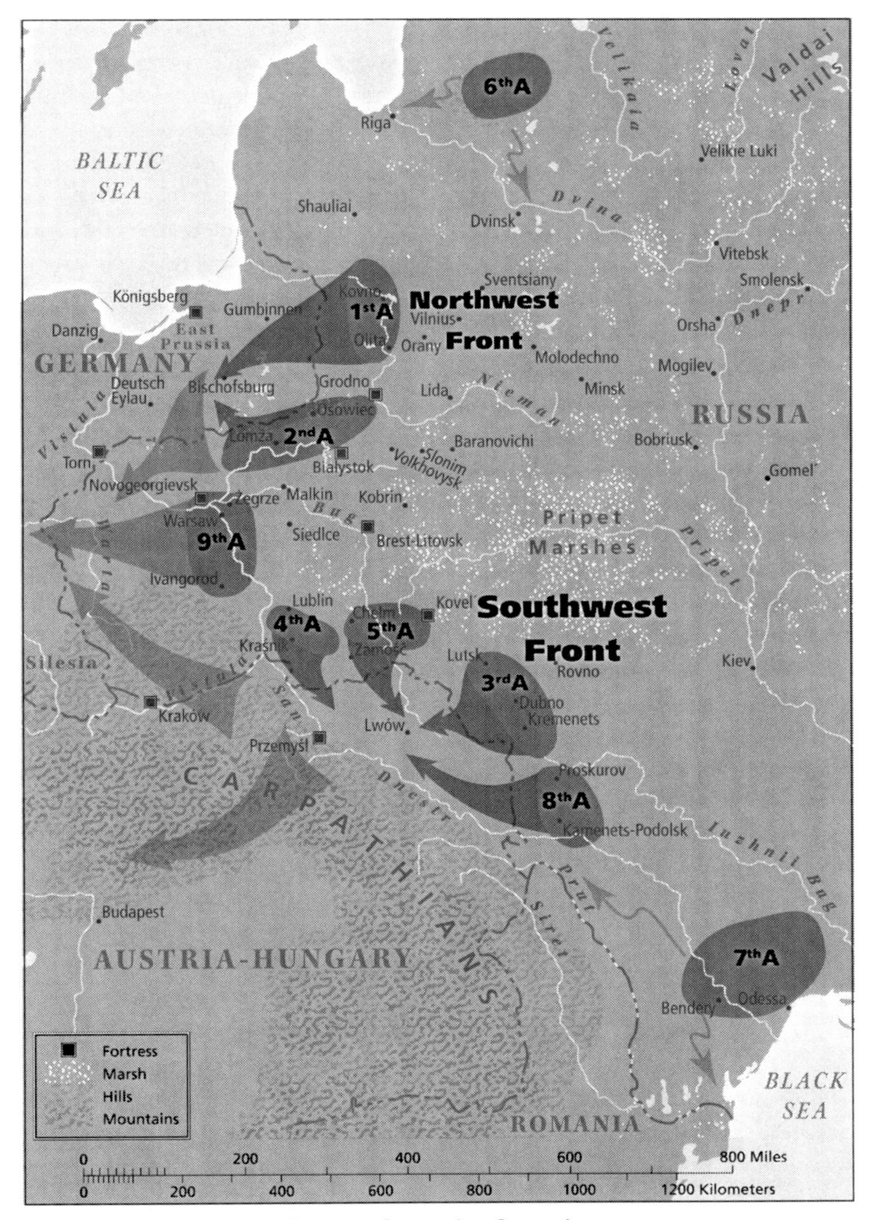

MAP 5. Concept: Successive Operations, 1914.

The result of various alterations in deployments, according to one former *genshtabist*, Ia. K. Tsikhovich, was that "in reality neither a true version of the 1 May 1912 plan nor a true version of the 25 September 1913 plan was implemented." Rather, as evident from alterations in deployments, the situation was governed "partially by one and partially by the other, along with such changes that were envisioned by neither."[113]

Last-minute improvisations were in large part a function of the way that key *GUGSh* personnel migrated from St. Petersburg to Baranovichi to form *Stavka*. Grand Duke Nicholas Nikolaevich's appointment as supreme commander was unexpected, and the tsar did not permit him to form his own field staff. To insure continuity from planning to execution, the tsar insisted that both *GUGSh* chief Ianushkevich and General Danilov, his chief of operations, occupy corresponding positions within *Stavka*. In concert with General Zhilinskii, the commander of Northwest Front, these officers were in a position to press their own priorities in favor of the anti-German effort and in stronger support of the alliance with the French. There was powerful sentiment that Russia owed France something in return for its unqualified support during the July Crisis that had preceded hostilities. Also, officers in high places had made promises that they were now honor-bound to keep. At the same time, Russian self-interest lobbied persuasively in favor of a combined effort to defeat the common enemy quickly; Germany must not be allowed to dispatch with the alliance partners *ad seriatim*. The 60-day window of opportunity loomed large in some quarters, especially within *Stavka*. The combined sense of obligation, urgency, and opportunity made up a powerful elixir. Under its influence, and in the absence of hard intelligence on the Germans, the Russians would fight the plan and their assumptions and not the enemy.[114]

Northwest Front: The Quest for Cannae

On Northwest Front, these factors, together with failures to attain mass and to exercise adequate command and control, would add up to a 1914 version of "the perfect storm." With its First and Second armies, the intent for Northwest Front was to conduct converging offensive operations respectively from the middle Niemen and the Narew to force a

[113] Tsikhovich (comp.), *Strategicheskii ocherk*, 20.

[114] Golovin, *Plan voiny*, 230–32; Skvortsov, Rushin et al. (eds.), *General'ny shtab Rossiiskoi armii*, 96–7; and Sukhomlinov, *Vospominaniia*, 308.

Cannae-style double envelopment on the defending German Eighth Army on either side of the Masurian Lakes. In the event the Germans refused decisive battle, the Russians would drive them westward across the lower Vistula and then either besiege or blockade Königsberg. As the creation of the Russian Ninth Army indicated, both *Stavka* and Northwest Front had already come to assume that the incursion into East Prussia was only a shaping operation. German defeat or withdrawal would trigger a regrouping of First Army assets on the middle Vistula. There, they would join forces with the Ninth Army to plunge across Silesia for a decisive operation in the direction of Berlin.[115] Within both *Stavka* and the headquarters of Northwest Front, the web of assumption and wishful thinking gained such credibility that it distorted perceptions of what was transpiring until the storm actually broke.

For their part, the Germans had long anticipated a Russian advance into East Prussia along several axes. The region was blessed with a dense railroad network and well-sited defensive strongholds and fortresses. Small as they were (8.5 divisions), German field forces counted a two-to-one edge over the Russians in heavy artillery firepower. The governing defensive concept, tested during repeated war games, was to engage in economy of force operations to defeat the Russians sequentially and in detail. Successful execution by General Max von Prittwitz's Eighth Army would require initiative, iron nerve, reasonably accurate intelligence, and the good sense to commit fewer mistakes than the Russians.[116]

Badgered to initiate offensive operations close to M+15 and before completion of concentration, Russia's First Army, led by General P. K. Rennenkampf, crossed the German frontier from the middle Niemen on 3 (16) August with a mere 6.5 under-strength infantry divisions and three cavalry groups. Russian aircraft rapidly succumbed to friendly fire and mechanical difficulties, so it was the mission of the cavalry groups to provide timely and accurate reconnaissance. However, they were invariably in the wrong place at the wrong time.[117]

Rennenkampf's army bumbled half-blind into an engagement just over the frontier at Stallupönen and a subsequent major battle on 6–7

[115] I. I. Vatsetis, *Operatsii na vostochnoi granitse Germanii v 1914. Ch. 1-aia. Vostochnaia-Prusskaia operatsiia* (Moscow-Leningrad, 1929), 27–8, 29n., 37–8; see also, Golovine, *The Russian Campaign of 1914*, 88–92.

[116] Terence Zuber, *German War Planning, 1891–1914: Sources and Interpretations* (Woodridge, Suffolk, 2004), 137–39, 144–45, 170–73; see also, Dennis E. Showalter, *Tannenberg, Clash of Empires* (Hamden, CT, 1991), 153–59.

[117] Vatsetis, *Operatsii*, 36, 52–3.

(19–20) August at Gumbinnen, both of which encounters bore characteristics of meeting engagements. In the first, elements of two Russian corps gradually overpowered Germany's lone defending I Corps. However, it liberally applied heavy artillery firepower to break contact and cover its withdrawal to the Angerapp, in the vicinity of Gumbinnen. There, during a major confrontation between the two armies, General von Prittwitz attempted a double envelopment of Russian dispositions on the Rominte River. He was successful in the north, using I Corps to overrun Rennenkampf's right-hand 28th Division, but suffered heavy losses during an attack against III Corps in the Russian center. There, during a series of vicious meeting, General August von Mackensen's XVII Corps outran its artillery support, and Russian 25th and 27th divisions met the attackers head-on with local artillery superiority and well-developed counter strokes. Mackensen's troops fled in disarray, while German forces in the south enjoyed only local success. Although neither side held absolute sway at dusk on 7 (20) August, it was the Germans who retired from the field, leaving behind only a screening cavalry division.[118]

The rout of German XVII Corps temporarily unnerved Prittwitz, who had simultaneously learned from aerial intelligence reports that General A. V. Samsonov's Russian Second Army had crossed the Narew to enter East Prussia from the southeast. With only his reinforced XX Corps defending in that sector, Prittwitz was wary of entrapment and envelopment between the two Russian pincers. Meanwhile, he also learned, from radio intercepts, that Rennenkampf's First Army was on a short tether because of supply and ammunition shortages. Recovering his nerve, Prittwitz bade his Eighth Army to regroup and redeploy against Samsonov, whose army was now deemed the more dangerous threat. Reinforced with local reserve and fortress troops, the reconstituted XVII Corps would move by forced march against Samsonov's right in the vicinity of Bischofsburg, while I Corps would entrain for deployment in the vicinity of Deutsch Eylau against Samsonov's left. If German XX Corps could hold in the center, Samsonov and not Prittwitz might fall into the deadly clasp of a double envelopment. However, Prittwitz would never bring his plan to fruition. His brief bout with nerves had called his competence into question, and the German high command replaced him

[118] Showalter, *Tannenberg*, 162–65,172–89; see also, Tsikhovich (comp.), *Strategicheskii ocherk*, 72–3; *Vostochno-Prusskaia operatsiia*, 90–2; and A. K. Kolenkovskii, *Manevrennyi period pervoi mirovoi imperialisticheskoi voiny 1914 g.* (Moscow, 1940), 184–85.

with General Paul von Hindenburg, who brought with him an able chief of staff, General Erich Ludendorff. Somewhat later, on 13 (26) August, the German high command made a second important decision. After the capitulation of the Belgian fortress at Namur, they dispatched 2.5 corps east to reinforce the beleaguered Eighth Army. Thus did initial Russian offensive operations serve immediate alliance interests.[119]

At the same time, the German retirement from Gumbinnen uncovered fundamental Russian weaknesses and played into flawed Russian assumptions. First Army might temporarily live off the land, but its troops were short of ammunition and casualty replacements. Reserve augmentation and logistics assets were still far down the mobilization and transit pipeline. First Army's three cavalry groups failed to track the regrouping German Eighth Army. In the absence of reliable intelligence, Rennenkampf and Zhilinskii, commander of Northwest Front, assumed that their opposition was withdrawing either to Königsberg or to the lower Vistula. The same assumption governed at *Stavka* and within the headquarters of Samsonov's Second Army. Both *Stavka* and Northwest Front now goaded the two Russian armies to step up the pace along their diverging directions of advance, with little heed to an opening gap between the two that would render mutual support impossible.[120]

Russian Second Army, meanwhile, displayed all the shortcomings of First Army – and more. Its staff had been hurriedly cobbled together, and its commander had only recently arrived via the Caucasus from distant Turkestan. Cavalry and communications assets, like logistics support, were scarce to non-existent. There was little intelligence on German dispositions, save for the vague notion from information on earlier German war games that the enemy might spring an envelopment from Deutsch Eylau in the West.[121] Perhaps worse, Second Army was less homogeneous than First, in part because XIII Corps came from outside the Warsaw military district. *Stavka* added to the muddle with another last-minute shuffling of assets that temporarily allocated I Corps (on the way to Ninth Army at Warsaw) to Samsonov's vulnerable left. The impulse to stretch initial deployments ever westward endowed Second Army with a wide frontage in an area astride the state frontier with few roads and only one potential supporting railroad (Mława-Soldau). Moreover,

[119] Showalter, *Tannenberg*, 191–99, 294–96.

[120] Melikov, *Strategicheskoe razvertyvanie*, 313; Golovine, *The Russian Campaign of 1914*, 156–59; and F. A. Khramov, *Vostochno-Prusskaia operatsiia 1914 g.* (Moscow, 1940), 41–2.

[121] Golovine, *The Russian Campaign of 1914*, 156–58, 181; and Golovin, *Plan voiny*, 92.

concentration areas were located farther from the state frontier than in First Army's sector, resulting in a lengthy march-maneuver over unfavorable terrain to initial enemy contact.[122]

In light of these disadvantages, the emphasis from above on speed became lethal. Reconnaissance from miscellaneous cavalry formations was poor, and after crossing the border on 7 (20) August with 9.5 infantry divisions, Second Army pressed forward half-blind across a broad front that became even broader as interventions from Northwest Front opened its flanks still farther to cover against possible German onslaughts from both the West and the Northeast. Stiff German resistance came only in the left center from XX Corps. Soon, Second Army's frontage extended nearly 100 kilometers across an area of East Prussia that was heavily forested and dotted with many lakes and swamps. Hindenburg and Ludendorff, meanwhile, were relatively well informed of Samsonov's extended dispositions, thanks to intercepts of radio messages transmitted *en clair*, intermittent but accurate aerial reconnaissance, and timely battlefield intelligence. As the redeployments initiated by Prittwitz focused initially on Second Army's exposed flanks, the momentum of its advance against German XX Corps in the center played into the enemy's operational design.[123]

The result between 14 (27) and 17 (30) August was the semi-annihilation of Second Army in the vicinity of the Kommusin Forest. Supported with heavy firepower that outranged Russian light batteries, German XVII and I Corps assaulted Russian Second Army's flanks, paring them away from its three center corps. Obeying his soldier's instinct to march to the sound of the guns, Samsonov on 15 (28) August abandoned his field headquarters, thereby losing the ability to command and control his army. Too late, he came to realize that he confronted not just a reinforced corps, but the entire German Eighth Army, in addition to various reserve and garrison formations. Except for General N. N. Martos, who commanded XV Corps in the Russian left center, Samsonov's corps commanders failed to take up the leadership slack, mostly for reasons of ineptitude, lassitude, and inexperience.[124]

[122] Vatsetis, *Operatsii*, 34–5.

[123] *Vostochno-Prusskaia operatsiia*, 281, 551–54; Rostunov, *Russkii front*, 120–21; Showalter, *Tannenberg*, 194–96, 228–34.

[124] Showalter, *Tannenberg*, 238–319; and Golovine, *The Russian Campaign of 1914*, 218–353. The report of the Russian investigative commission on the demise of the Second Army is reprinted in *Vostochno-Prusskaia operatsiia*, 547–60.

For its part, Northwest Front headquarters remained a half-ignorant bystander, its actions either ineffectual or late. And its most important asset, First Army, was now too distant to render assistance.[125] *Stavka*, meanwhile, suddenly became preoccupied with the possible annihilation of two entire armies on the right flank of Southwest Front.[126] Another impending disaster some 300 kilometers distant from the Second Army forced the Headquarters of the Supreme Commander to send the remaining elements of the still-forming Ninth Army not to succor Second Army, but to the southeast to forestall a second disaster.[127] Although the shell-shocked corps on Second Army's flanks managed to elude the German grinding mill, Hindenburg was able to effect a shallow envelopment of Samsonov's center, nearly obliterating three entire Russian corps (XV, XIII, and elements of XXIII). In all, the victory that the Germans would call Tannenberg cost the Russians approximately 90,000 casualties and prisoners of war, the latter including two corps commanders.[128] Samsonov took his own life on the night of 16–17 (29–30) August.

Failures in anticipation, mutual support and command and control continued to plague the Russian effort in East Prussia. After Gumbinnen, First Army had maintained a leisurely-paced advance westward, with the intention of securing the lower Vistula and besieging Königsberg. After Tannenberg, Hindenburg and Ludendorff regrouped their forces by foot and rail to pin Russian First Army frontally and envelop its exposed southern flank. The result between 24 August (6 September) and 1 (13) September was a Russian withdrawal from East Prussia by leaps and bounds, with occasional savage confrontations for key defensive lines and transportation junctions. In retreat, Rennenkampf committed his forces piecemeal and often without sufficient coordination, so that his losses mounted, first gradually and then precipitously. However, he counted two unexpected advantages, and they were crucial to First Army's survival. First, despite the arrival of 2.5 fresh corps from France,

125 There is no substance to the legend that Rennenkampf allegedly failed to aid Samsonov because of animosity between the two army commanders dating to a physical altercation between them at Mukden during the Russo-Japanese War. See, Showalter, *Tannenberg*, 134.

126 At *Stavka*, General Danilov's telling comment was: "The absence of communications with Samsonov is serious, but after all, he has five corps, and our failure there would scarcely have decisive significance." See, *Vostochno-Prusskaia operatsiia*, 298; and Vatsetis, *Operatsii*, 146, 182–83.

127 *Vostochno-Prusskaia operatsiia*, 18–19, 364.

128 Ibid., 316–18, 320.

Hindenburg's troops were nearing exhaustion. Their pursuit often fell short of relentless. And, second, Hindenburg and Ludendorff temporarily lost their operational touch, choosing to pursue Russian First Army across a broad front, rather than attaining mass on either vulnerable Russian flank. By the end of the second week of September (N. S.), First Army had retreated to its original starting point behind the Niemen, but with casualties nearly matching those of Second Army.[129]

The Russian War at Sea

While these events transpired in the land war, the Russian navy discharged its functions as "the second arm." At the outbreak of hostilities, Admiral A. A. Ebergard's Black Sea Fleet remained concentrated at Sevastopol. The arrival of the German battlecruiser *Göben* and light cruiser *Breslau* at Constantinople signaled Turkey's possible entry into the conflict, but Ebergard had no operational plan for this eventuality. In accordance with directives, the Black Sea Fleet during the first month of the war limited its actions to deploying a barrier of 330 mines to protect the naval roadstead at Sevastopol.[130]

In the north, the Baltic Fleet mined the approaches to its primary bases on 15 (28) July, and on 18 (31) July) laid the barrier (2,124 mines in eight belts) for the "central position" in the Gulf of Finland. On 20 July (2 August), an additional 200 mines were sown to cover the "flank position" between Sveaborg and the Skerries. The Russian naval bases at Libau and Vindau were evacuated and their entrances blocked by sunken cargo vessels. On 20 July (2 August) to 22 July (4 August), two German light cruisers laid mines at the harbor entrance to Libau and shelled the city. On 27 July (9 August), the Russian battle squadron sallied against German cruisers, but was recalled. German cruisers remained active, laying several hundred mines off Hanko. Between 10 (23) and 14 (27) August, German cruisers even threaded their way through the Russian mine barrier to enter the Gulf of Finland. However, the *Magdeburg* ran aground off Odensholm, and Admiral N. O. Essen hastened to the scene with two Russian cruisers. Although the Germans had abandoned ship

[129] Golovine, *The Russian Campaign of 1914*, 354–90, 373–82; Khramov, *Vostochno-Prusskaia operatsiia 1914 g.*, 94–6; Tsikhovich (comp.), *Strategicheskii ocherk*, 111–13; and Showalter, *Tannenberg*, 325–26.

[130] Zolotarev, "Deistviia russkogo flota v Pervoi mirovoi voine," 56–7.

and set demolition charges, only the bow was blown off, and Russian divers sent below to inspect the damage soon came up with the German naval code books, which had been thrown overboard. These materials the Russians eventually shared with the British and French, enabling them to decipher the German naval code throughout World War I.[131]

In general, Admiral Essen wanted to challenge German supremacy on the Baltic, but the high command restrained him. On 9 (22) September, Essen learned that the Germans were to land a token expeditionary force on the Russian Baltic coast to divert attention from operations against the Austro-Hungarians in Galicia. Essen took his battle squadron to the mouth of the Finnish Gulf, but *Stavka* ordered his return. *Stavka* never approved his subsequent requests to operate against German lines of communication in the Baltic. Even after the four dreadnoughts for the Baltic Fleet were launched during the closing months of 1914, they were not permitted to sail without the express permission of the tsar.[132] Meanwhile, the funds for their construction were sorely missed on Southwest Front, where only the dilapidated fortress at Ivangorod protected the exposed Russian right flank.

Southwest Front: Neither Cannae nor Leuctra

General N. Iu. Ivanov's Southwest Front confronted the same requirements as Northwest Front for speed and adaptation on the run. But his more seasoned command and staff cadres overcame odds to weather the unexpected, only to fall prey to the same iron laws of mass and logistics that had so hamstrung the two Russian armies in East Prussia. Unlike the situation on Northwest Front, miscalculation based on prewar intelligence estimates initially played a major role in confrontations against the Austro-Hungarians. And, unlike the closed terrain in much of East Prussia, the open expanses of Galicia held great promise for success in the kind of maneuver warfare that Manchuria had taught the Russians to expect. However, there would be neither the Leuctra for which Alekseev had planned, nor the Cannae that Danilov had anticipated.

Speed assumed the same overriding role as on Northwest Front. Alliance obligations figured only indirectly in the operational calculus

[131] F. N. Gromov and I. V. Kasatonov (eds.), *Na rubezhe vekov: Tri veka Rossiiskogo flota*, 3 vols. (St. Petersburg, 1996), III, 69–71.

[132] Ibid., 72–3.

for Southwest Front, but two major factors contributed to the greater emphasis on speed. The first was the understanding that Austria-Hungary might undertake offensive operations as early as M+16. The second was raw self-interest. If Germany might administer a *coup de grace* to France in approximately two months, formations from Southwest Front would be required to reinforce the general anti-German effort, either offensively or defensively, across the middle and lower Vistula. As on Northwest Front, the obsession with speed played havoc with notions of mass and concentration. Also, thanks to Danilov's tinkering with deployments in 1913 and the recent redeployment of XX Corps to Northwest Front, there was no heavy right-flank hammer to deliver a Leuctra-like blow from the northwest to Austro-Hungarian dispositions in Galicia. Now, just as in East Prussia, the concept was to engage the Austro-Hungarians in a gigantic reincarnation of Cannae, possibly north of Lwów. To this end, and before completion of concentration, the two right-flank armies (the Fourth and Fifth) of Southwest Front would receive orders to initiate march-maneuver to contact on 10 (23) August. Although the Third and the Eighth armies on the left flank would be more fully concentrated, they would be pressed into offensive operations even earlier (5 [18] and 6 [19] August respectively), in part because they had more distance to cover, and in part because they were to distract enemy attention from the vulnerable Russian right south of Lublin and Chełm.[133]

Unfortunately for the Russians, the Austro-Hungarian Chief of the General Staff, Franz Conrad von Hötzendorf, was of no mind to be steam-rollered. After a year out of favor, he had returned to the General Staff in late 1912. He understood the precarious nature of initial Habsburg wartime dispositions in Galicia, and he knew that the Russians possessed hard 1912-vintage intelligence on those dispositions. He also understood that the fragile nature of his multi-ethnic army required immediate offensive success. For Conrad, the best prospect for that success would come from a knockout blow in conjunction with the Germans against Russian dispositions at Siedlce and Brest-Litovsk. Therefore, in late 1913 and early 1914, Conrad altered Austro-Hungarian deployments in Galicia. Unknown to the Russians, he shifted them 40–100 kilometers to the West and reinforced his western flank so that it comprised two heavy

[133] Norman Stone, *The Eastern Front, 1914–1917* (New York, 1975), 82, 84; A. M. Zaionchkovskii, *Pervaia mirovaia voina*, reprint ed. (St. Petersburg, 2000), 96–7; Golovin, *Galitsiiskaia bitva*, 11–12, 63–5, 73–5; and Tsikhovich (comp.), *Strategicheskii ocherk*, 54–6.

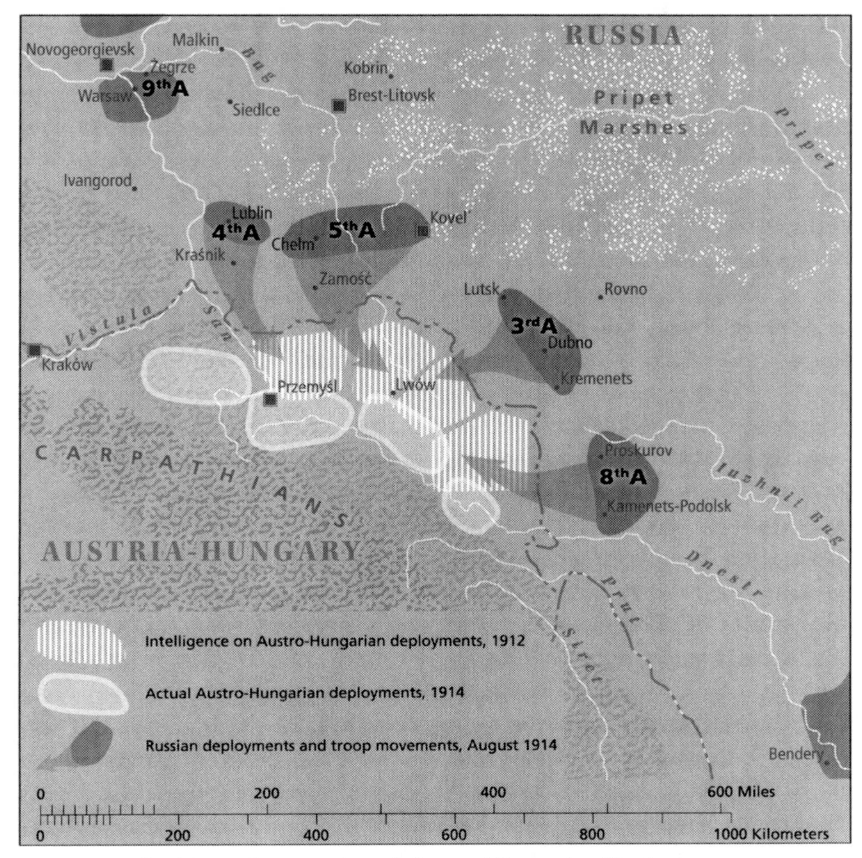

MAP 6. Anticipated versus Actual Austro-Hungarian Deployments, 1914.

shock armies (the First and Fourth) deployed between the Bug and the Vistula.[134] These alterations meant that Russian Southwest Front's two right-hand armies would advance not into the soft flank of Conrad's deployments north of Lwów and Przemyśl, but into the teeth of a major Austro-Hungarian offensive operation. As on Northwest Front, Russian aerial and mounted reconnaissance assets – in some cases far back in

[134] Günther Kronenbitter, *"Krieg im Frieden": Die Führung der k.u.k. Armee und die Großmachtpolitik Österreich-Ungarns 1906–1914* (Munich, 2003), 446–47; see also, Graydon A. Tunstall, Jr., *Planning for War against Russia and Serbia: Austro-Hungarian and German Military Strategies, 1871–1914* (New York, 1993), 96; [Borisov], *Kratkii strategicheskii ocherk voiny*, I, 36–7.

the mobilization and transit schedules – failed utterly to provide updated intelligence on enemy dispositions.[135]

During August 1914, changed enemy dispositions, together with the time-distance calculus governing the closure rate from east of the Russian Third and Eighth armies, conditioned the content and outcome of initial operations on the Southwest Front. For the Russians, these operations in the West quickly devolved from a march-maneuver to contact into a desperate race between adaptation and catastrophe. As early as (9) 22 August, *Stavka* had warned General Alekseev, chief of staff for Southwest Front, that the Austro-Hungarians might have altered their dispositions. In response, Alekseev modified the direction of attack for Ivanov's two right-hand armies and echeloned their fronts to confront potentially changed enemy deployments more directly.[136] On the following day, Fourth Army (6.5 divisions) commenced the advance on the extreme Russian right. Roughly 30 kilometers southwest of Lublin, it ran head-on into 10.5 divisions from General Viktor Dankl's Austro-Hungarian First Army. XIV Corps on the Russian right was severely mauled in a meeting engagement, and two hard days' fighting ensued in the vicinity of Kraśnik, with the tactical advantage changing hands several times. Finally, with both flanks threatened, the Russian Fourth Army withdrew to its original jumping off positions, and the call went out for reinforcements. Adapting to the situation, Alekseev's response was to wheel General P. A. Pleve's neighboring Fifth Army (8.0 divisions) to the Southwest to outflank and cut through the rear of Austro-Hungarian First Army. However, Russian Fifth Army collided with General Moritz von Auffenberg's reinforced Fourth Army (10.5 divisions), and meeting engagements soon exploded across Pleve's center. Outnumbered, the Russian Fourth and Fifth armies now fought only to retain terrain and local advantage, and what had begun for the Russians as an exercise in pinning the enemy and maneuvering for his open flank degenerated into a series of close-range slug fights. As Austro-Hungarian numbers exacted their toll, both Russian armies recoiled, yielding local victories to the Austro-Hungarians at Kraśnik and Zamość-Komarów.[137]

With their backs against the Lublin-Chełm railroad, the Russians sought salvation in dogged defensive action, continued adaptation, and

[135] Golovin, *Galitsiiskaia bitva*, 353; Kolenkovskii, *Manevrennyi period*, 105, 225.

[136] Tsikhovich (comp.), *Strategicheskii ocherk*, 47; [Borisov] *Kratkii strategecheskii ocherk*, I, 36–8, 137–38.

[137] Kolenkovskii, *Manevrennyi period*, 223–24; Beloi, *Galitsiiskaia bitva*, 89–90; and Rostunov, *Russkii front*, 131–36.

the slow buildup of numbers. For reasons that remain unclear, corps commanders on the Southwest Front were generally more capable and experienced than those on the Northwest Front. For example, XIX Corps commander, V. N. Gorbatovskii, was a hardened veteran from the defense of Port Arthur in 1904, and he was able to fight his way out of a near-encirclement. Several less competent commanders were relieved. Meanwhile, after consulting with *Stavka*, Alekseev and Ivanov altered dispositions, even permitting army-level commanders the luxury of withdrawal to reconstitute defensive positions and to reduce frontages. Brigade-level formations were shifted to danger spots and reserve formations were fed into battle as they debarked from their railroad cars.[138] However, the half-ruined fortress at Ivangorod was in no condition to anchor the increasingly shaky western flank. Therefore, on 17 (30) August, in response to urgent requests from Alekseev, *Stavka* released reinforcements from the still forming Ninth Army to shore up Ivanov's right.[139] Even as Russian Second Army was in its death rattle, the anti-German emphasis at *Stavka* suddenly became less important. The additional numbers would have telling effect on Southwest Front, but full redeployment would require nearly a week.

In the interim, Alekseev attempted to recapture his vision of Leuctra – this time from the East – but his effort came to naught, thanks to a failure in command and control. While his right-flank armies were fighting for their lives south of Lublin and Chełm, his two left-flank armies, the Third and Eighth, were marching through a series of victories against varying degrees of opposition east of Lwów on the Złota Lipa and Gniła Lipa (13–16 [26–29] August).[140] Conrad had not anticipated the power of this onslaught from Podolia and Volhynia, and he had starved his secondary effort in the East to feed likely success in the West. Timely reinforcement from troops originally destined for the Serbian front was slow to come. The result was the imminent loss of Lwów, Galicia's capital city. Indeed, General N. V. Ruzskii, commander of the Russian Third Army, now engaged in his own version of fighting the plan rather than the enemy. Ruzskii remained so fixated on the seizure of Lwów that he and his staff ignored repeated orders from Southwest Front headquarters to shift the axis of his advance to the right and to extend his right flank to the Northwest. To do so would have closed the gap between the Third and

[138] Rostunov, *Russkii front*, 137–40; Golovin, *Galitsiiskaia bitva*, 367–440.
[139] Golovin, *Galitsiiskaia bitva*, 367–68; Stone, *The Eastern Front*, 85–7.
[140] Stone, *The Eastern Front*, 88–9; Rostunov, *Russkii front*, 140–45.

Fifth armies and positioned Third Army to roll over the rear of the two Austro-Hungarian armies threatening the entire right flank of Southwest Front.[141] It was not until Ruzskii had occupied Lwów on 21 August (3 September) and Grand Duke Nicholas Nikolaevich had personally intervened that Third Army fell fully into the operational fold, by which time the crisis on the Russian right had subsided.[142] Reinforcements from the Ninth Army had begun to turn the tide south of Lublin and Chełm, although by now Alekseev had lost his opportunity for a Leuctra-like battle of annihilation. However, there were now enough numbers to confront Conrad directly in all sectors. On the same day that Third Army took Lwów, Ivanov bade his entire Southwest Front to shift to the counteroffensive.

In response, Conrad withdrew his two left-hand shock armies to shorten his lines, wheeling Fourth Army to the Southeast to confront the threat from Russian Third Army. In addition, he threw divisions transiting from the Serbian front into a last-gasp counteroffensive west of Lwów and called for the Germans to attack from East Prussia south toward Siedlce to relieve pressure from the North. However, the German high command refused, since it rightly judged the moment inopportune until Eighth Army had cleared East Prussia of withdrawing Russian First Army. By the time that mission was complete in mid-September, Conrad had retired across the San, and advancing troops from Russian Southwest Front were threatening his communications and deployments in the vicinity of the fortress at Przemyśl.[143]

It was at this point that Russian operations against the Austro-Hungarians began to culminate from a combination of casualties, exhaustion, and logistical over-extension. While on the defensive, the Fourth and Fifth armies had their backs against supporting railroads. Following transition to the counteroffensive, the same armies, now reinforced, lacked logistical support to operate beyond three to four days' marching distance from their railheads. Worse, small arms munitions and artillery shells were running short, and there was no re-supply in sight.[144] Still worse,

[141] Golovin, *Galitsiiskaia bitva*, 60, 115–194-97, 328–30, 348, 510; see also, Stone, *The Eastern Front*, 88.

[142] Golovin, *Galitsiiskaia bitva*, 351, 363.

[143] Tunstall, *Planning for War against Russia and Serbia*, 242, 245–46, 251–52.

[144] See, for example, General'nyi Shtab RKKA, *Varshavsko-Ivangorodskaia operatsiia. Sbornik dokumentov* (Moscow-Leningrad, 1938), 445–47; and the overview in Beloi, *Galitsiiskaia bitva*, 108–113, 140.

on the average, Russian divisions had suffered 40 percent casualties over nearly three weeks' hard fighting. Total casualties numbered about a quarter million, and replacements came in dribbles, with fresh formations still far down the list on the mobilization and rail transit tables.[145]

Notable success had followed on the heels of initial travail, but Russian offensive operations on Southwest Front stalled out to the tune of plaintive calls for supplies and reinforcements. By the time these arrived in any appreciable quantities, Conrad would have recovered his equilibrium, and the Russians would be confronted not just by the remnants of Conrad's field armies, but also by German reinforcements from East Prussia.[146] Their appearance west and southwest of Warsaw would necessitate a regrouping of Southwest Front's armies to confront an ominous threat in an altered locale. During the initial period of war, Russian operations on Southwest Front had spent their force ("culminated," in military parlance) short of decision, with the result that the fight – just as on Northwest Front – would become protracted.

Conclusion

Writing in France more than a decade after the fact, General Golovin blamed various Russian failures in August 1914 on the war plan, even as the term was loosely construed at the time.[147] However, it is military axiom that "no plan survives initial contact." Indeed, Helmuth von Moltke (the Elder), the chief of the Prussian General Staff often credited as the godfather of modern war planning, held that "no plan of operations extends with certainty beyond the first encounter with the enemy's main strength." Yet, Moltke did not discount the importance of the plan in shaping subsequent operations although he also held that "a mistake in the original assembly of the army can scarcely be rectified in the entire course of the campaign."[148] After 1905, Russian preparation and planning for a European war admitted numerous mistakes in anticipation of initial operations, with many errors emanating from systemic-oriented lapses in collaboration, calculation, and competence.

145 Zaionchkovskii, *Pervaia mirovaia voina*, 259.
146 Showalter, *Tannenberg*, 326–27.
147 In *The Russian Campaign of 1914*, 175, Golovin[e] called the Russian war plan "criminally thoughtless."
148 Daniel J. Hughes (ed.), *Moltke on the Art of War: Selected Writings*, tr. Daniel J. Hughes and Harry Bell (Novato, CA, 1993), 45.

Once various mistakes condemned the Russians to poor opening moves at war's beginning, commanders and staffs in the field would be thrown back on their capacities to learn and adapt on the run, under combat conditions. But there was little in August 1914 to suggest that they might consistently rise to the task. Chess players with poor opening moves seldom win tournaments.

5

France

Robert A. Doughty

In the years following its defeat in the Franco-Prussian War, France entered a new era in its military history and its strategic planning. France had gone to war in 1870 with little forethought and preparation, and its swift defeat came not only from the excellence of the Prussian military but also from its own confusion and mistakes on the battlefield. In 1874, a committee in the National Assembly charged with proposing a new law on the administration of the army stated, "[W]e cannot stamp our foot upon the ground and expect invincible armies to emerge. . . . [W]e must prepare and organize in advance the armed forces of the nation if we want them, at the time of danger, to be ready for prompt and energetic action."[1] Over the next forty years the French attempted to think through the military challenges they could face and slowly but surely adapted to the contingencies and threats they perceived. By August 1914, they had thoroughly reformed their army, but the opening battles of World War I revealed many shortcomings. While planning and preparation had helped, errors in strategy, weapons acquisition, and doctrine cost them dearly.[2] In the end, the ability to improvise enabled the French to win the Battle of the Marne, but the failure to anticipate and prepare for the long, deadly war that ensued cost them even more.

[1] *Journal Officiel de la République Française* (1874) (hereafter *J.O.*), p. 5718.

[2] On the general topic of planning, see Thomas Adriance, *The Last Gaiter Button: A Study of the Mobilization and Concentration of the French Army in the War of 1870* (New York, 1997); Gary P. Cox, *The Halt in the Mud: French Strategic Planning from Waterloo to Sedan* (Boulder, 1994); and A. Marchand, *Plans de concentration de 1871 à 1914* (Paris, 1926).

Perceived Threats and National Goals

In the wake of the 1871 debacle, France did not have to search far for threats. Foremost among these was the newly unified power in central Europe. As Germany's population and economic power expanded and overshadowed those of France, it dominated French military thinking. In the age of new imperialism French soldiers and sailors carried the tricolor around the globe, but they never took their eyes off Germany. Political and military leaders measured the readiness of the army by comparing the numbers and qualities of their soldiers and weapons with those of the Germans. They compared defense budgets and paid close attention to the amount spent per soldier. They also watched carefully what the Germans did with new technologies in areas such as aviation and communications.

Several war scares in the 1870s and 1880s heightened France's feeling of vulnerability. Then, beginning with the first Moroccan crisis of 1905, a series of diplomatic trials of strength worsened international tensions and increased the chances of war. In the Moroccan crisis, political and military leaders recognized that France almost certainly would lose if it went to war with Germany and that its allies were unwilling and incapable of offering any help. Over the next decade the threat of a war with Germany increased but became especially grave after the second Moroccan crisis of 1911 and the Balkan wars of 1911–12. When Germany responded to these crises with an expansion of its army and the acquisition of new armaments, very few French leaders doubted that the Germans posed a significant threat. A formal assessment in 1911 of the "most likely conflict" facing France stated: "The General Staff of the Army considers it obvious that Germany is our principal adversary.... [O]f all the rival nations for France, Germany is the strongest, the most immediately menacing. Its allies are only satellites."[3]

The threat from Germany did not translate into broad public support for a preventive war or a war of revenge. For decades after the Franco-Prussian War anger at the defeat and the loss of Alsace and Lorraine had undermined any chance of friendly relations between the two states but had not led to war. In 1882, Paul Déroulède's League of Patriots contributed to an upsurge of chauvinism and a call for a return of the two lost provinces. General Georges Boulanger, who served as the minister of war in 1886–88, became known as "General Revenge" for his symbolizing a

[3] M.G., É.M.A., Note indiquant les points relatifs à la situation extérieure, octobre 1911, Service historique de la Défense (hereafter SHD) 2N1, p. 2.

more aggressive policy toward Germany. Later, Raymond Poincaré, who served as President of the Council of Ministers (Premier) and President of the Third Republic, became the spokesman of the nationalist revival of 1911–14 and increased firmness toward Germany. The French people, however, had grown less interested in revenge as the trauma of 1870–71 became more distant. Despite increased tensions after 1911, legislative elections in April and May 1914 seemed to indicate the French people's reluctance to support three years of military service or an aggressive foreign policy. Whatever threat Germany posed, France had to let Germany take the first step toward war.

France also had to pay attention to Italy, the newly unified state on its southeastern frontier. When the French withdrew troops from Rome during the Franco-Prussian War, the Italians capitalized on the defeat of France at Sedan and seized Rome. Over the next several decades the "Roman Question" remained sensitive, and conservative Catholics often demanded the freeing of the Papacy from the "tyranny" of the Italians. Though French political leaders had no intention of fighting a war with Italy to free the pope, they often complained of the Papacy's failure to cater to French public opinion and to put pressure on the Italians. Additional friction came from the clash of imperial ambitions in North Africa, especially after the French seizure of Tunisia in 1881. Shortly afterward, the Italians completed agreements with Vienna and Berlin and signed the first treaty of the Triple Alliance, which pledged German and Austrian assistance in the event of a French attack. Although a Franco-Italian tariff war that began in 1888 lasted for ten years, relations between France and Italy slowly improved, and by 1902 France expected Italy to remain neutral if France were attacked by another power. Relations improved even more when Italy failed to support its ally, Germany, at the Algeciras Conference of 1905 and when events in the Balkans made Italy uneasy about Austria-Hungary and more willing to befriend Russia and France. The Moroccan crisis of 1911, however, worsened relations, especially after the Italians seized Tripoli. In the rush of events in July 1914, French leaders were relieved when they received word Italy would remain neutral. Sizeable French forces had long occupied defensive positions along the Italian frontier, and Italy's declaration of neutrality freed them for use against Germany.

To respond to the German threat, French diplomats energetically pursued better relations with the Russians and British. They knew France could not face Germany alone. For strategic planners the advantages of an alliance with Russia were obvious, especially threatening Germany with a two-front war and forcing it to divide its forces and attention between two fronts. Though the diplomatic maneuvers of Otto von Bismarck kept France isolated for two decades after the Franco-Prussian War, an opportunity for closer relations between France and Russia appeared in March 1890 when Bismarck was forced out of office. Prior to this, contacts had been limited to such things as Russia's buying 500,000 Lebel rifles, but the tsar's doffing of his cap on the playing of *La Marseillaise* when the French fleet visited Kronstadt in July 1891 signaled the beginning of a very different relationship. After reaching an initial agreement a month later, the two countries completed a military convention in late 1892 and formally acknowledged it in an exchange of notes in December 1893 and January 1894.

The convention included the strategic concept that henceforth provided the foundation for all French planning: "Germany has to fight at the same time in the east and the west."[4] Russia would attack Germany if it attacked France or if Italy supported by Germany attacked France; France would attack Germany if it attacked Russia or if Austria-Hungary supported by Germany attacked Russia. The convention also included requirements for the general staffs of the two armies to plan future operations, coordinate the actions of the two armies, and establish effective communications between them. French strategic planners believed the Germans could not prevail on both fronts and, even if France or Russia were beaten, the other would prevail and assure the eventual victory of the two allies.[5]

Formal staff talks occurred in 1900, 1901, 1906, 1907, and 1908. In 1910 the two powers affirmed the discussions of previous meetings and emphasized, "Defeat of the German armies remains, no matter what the circumstances, the first and principal objective of the allied armies." The generals representing the two countries agreed that Germany probably would direct its main effort against France at the beginning of a war and send only small forces against Russia. The French representative emphasized France's intention to launch an "all-out and immediate" offensive against Germany at the beginning of a war.[6] Subsequent talks in 1911, 1912 and 1913 confirmed Germany's defeat as the "first and principal

[4] A copy of the convention appears in Raymond Poincaré, *Au service de la France*, vol. 1, *Le lendemain d'Agadir* (Paris, 1926), p. 292.

[5] Le lieutenant-colonel Pellé au général Brun, 24 mars 1910, *Documents diplomatiques francais* (hereafter *D.D.F.*), 2e série, vol. 12, annex no. 467, p. 717.

[6] Procès-verbal de l'entretien entre les chefs d'état-major généraux des armées française et russe, 7/20 et 8/21 septembre 1910, *D.D.F.*, 2e série, vol. 12, annex no. 573, pp. 911–13.

objective" of France and Russia and the defeat of Germany's allies as subordinate objectives. In each of the staff talks from 1910 to 1913 military representatives of the two countries discussed the word "defensive" in the convention's preamble and agreed that the word did not rule out the possibility of offensive operations. Additionally, the meetings reinforced the requirement for rapid mobilization and offensives by France and Russia.

In August 1913 the Russians agreed to begin an offensive on the fourteenth day after mobilization, and both allies promised to send their offensives into the "heart" of Germany.[7] The French representative, General Joseph Joffre, promised France would engage nearly all its forces on the northeastern frontier and commence offensive operations on the eleventh day after mobilization.[8] By July 1913, wireless communications between France and Russia were open each day between 0600 and 0800 and 2000 and 2400 hours, and backup links along an alternate route were open each night between midnight and 0200 hours.[9] These communication links ensured that French and Russian plans and actions could be tightly coordinated.

The relationship with the British was far more uncertain. Even though France had long courted Britain and in April 1904 signed the Anglo-French *entente*, the agreement did little more than liquidate outstanding colonial differences. Not until later did discussions between army and naval staffs increase the significance of what became known as the *entente cordiale*. From the earliest years of the *entente*, the British appeared to be more willing to provide naval forces, and their commitment of land forces remained uncertain. By July 1911, the French expected the landing of six infantry divisions and two brigades of cavalry, totaling some 150,000 men and 67,000 horses.[10] The French did not expect the British to add considerably to their fighting power; they simply expected the combined number of French and British forces to be "nearly equal" to those of Germany in a campaign in northern France and Belgium. They also knew the British could be delayed for political reasons or by transportation

[7] Renseignements données par le Général Gilinski au Général Joffre dans la conférence préliminaire du 30 juillet-12 août 1913, SHD 7N1535; P.V. des entretiens du mois d'août 1913, 24 août 1913, *D.D.F.*, 3e série, vol. 8, annex no. 79, p. 88.

[8] P.V. des entretiens du mois d'août 1913, 24 août 1913, *D.D.F.*, 3e série, vol. 3, annex no. 79, pp. 88, 90.

[9] É.M.A., Communications entre la Russie & la France (Traduction de la note russe), Mars 1911; Communications franco-russes, juillet 1913. Both are in SHD 7N1538.

[10] Memorandum de la conférence du 20 juillet 1911, p. 2, SHD 7N1782.

difficulties.[11] Until Britain declared war on Germany and began moving troops in August 1914, the French had no guarantee of British support. Additionally, the secrecy and sensitivity of conversations between Paris and London prevented any open mention of British assistance. Having Britain as an ally thus added to the number of soldiers who could participate in a campaign against Germany; it also enabled the French to concentrate their fleet in the Mediterranean and move troops more readily from Africa. The French, however, relied more heavily on and expected much more of the Russians.

The question of Belgium complicated planning significantly, for France's political leaders refused to allow any violation of Belgium's neutrality before the Germans did so. They insisted that Belgium would never agree to France's violating its neutrality and that such an action would probably result in its joining Germany. In a meeting of the Superior Council of Defense in October 1911, Joffre summarized the problem of Belgium's neutrality: "If we violate Belgian neutrality first, we will become the aggressors. England will not join our side; Italy will have the right to declare against us...."[12] In 1912 diplomatic and military officials discreetly probed the willingness of the British to countenance a violation of Belgium's neutrality, but blunt advice from General Sir Henry Wilson, the British director of military operations, ended these efforts. Strong opposition from political leaders and the British to violating Belgium's neutrality left military leaders no choice but to accept that neutrality. Even if the Franco-Russian alliance demanded immediate action, the French would not enter Belgium before the Germans.

Above all, French political leaders did not wish to be saddled with the responsibility for starting a war with Germany. They knew the French people did not desire war and would not support one if they thought their political leaders had sought war or had manipulated events to cause one. They recognized that the army could do little if large numbers of conscripted soldiers believed France was responsible for starting the war and refused to report for duty upon mobilization.

[11] 2e Bureau, Comparison des forces qui pourraient se trouver en presence sur le théatre du Nord-Est, 19 octobre 1912, SHD 2N1; M.G., É.M.A., 4e Bureau, Note pour l'état-major de l'armée (3e bureau), 4 mars 1913, SHD 7N1782; Séance du 4 juillet 1919, in France, Assemblée nationale, Chambre des deputés, Onzième législature, Session de 1919, vol. 71, no. 6206 (annexe), *Procès-verbaux de la commission d'enquête sur le rôle et la situation de la métallurgie en France* (*Défense du bassin de Briey*, 2e partie) (Paris, 1919), p. 159.

[12] P.V., C.S.D.N., Séance du 11 octobre 1911, Fonds Messimy, Archives nationales (hereafter AN) 509 AP/5.

Following the assassination of Archduke Francis Ferdinand, the Council of Ministers met on 30 July to discuss the rapidly worsening crisis. Though no formal minutes exist, Abel Ferry, the undersecretary of state for foreign affairs, recorded the main points of the meeting: "For the sake of public opinion, let the Germans put themselves in the wrong." He also noted, "Do not stop Russian mobilization. Mobilize, but do not concentrate."[13] France, clearly, did not offer Russia a "blank check," but it took only very modest steps to defuse the Balkan crisis of July 1914 or to restrain Russia.

Planners and Planning in the Third Republic

In the aftermath of 1870–71, the Third Republic's political leaders recognized that France needed skilled, professional officers to complete the complex task of thinking through and preparing for the many challenges of mobilizing, organizing, and employing a large force in war. Yet, experiences under Napoleon III and doubts concerning the political reliability of the army created significant concerns about creating a formal military hierarchy with real authority. Initially, the General Staff was little more than the military cabinet of the minister, but in 1874 the French enlarged and renamed it the General Staff of the Minister of War, thereby indicating not only its expanded role but also its subordination to the minister. Wary political leaders at first refused to follow the German model and designate a chief of staff, but they finally agreed to do so and named him Chief of Staff of the Minister of War. This title made it clear that he had no authority over the army and was subordinate to the minister. In 1890 Charles de Freycinet, the first civilian minister of war, changed the names to General Staff of the Army and Chief of Staff of the Army. Republican leaders nonetheless remained reluctant to appoint a single general officer head of the army during peacetime or create a strong military hierarchy or command structure.

The French partially compensated for the fragmented military hierarchy by establishing the Superior Council of War and the Superior Council of National Defense. The former, created in 1872, had the president as its chair, and included the premier and the army's senior generals. The council was supposed to be consulted "in a general manner" on all measures able to affect the constitution of the army and the manner in which it

[13] Note, *D.D.F.*, 3ᵉ série, vol. 11, p. 262n. Also note by Abel Ferry, 30 July 1914, Ministère des affaires étrangères, Archives privés de Abel Ferry; quoted in J. F. V. Keiger, *Raymond Poincaré* (Cambridge, 1997), p. 175.

would be employed.[14] The Superior Council of National Defense, formed by a decree on 3 April 1906, had a much broader task. With the president of the Third Republic presiding, the council included the premier and the ministers of war, navy, foreign affairs, finance, and colonies. The ranking officers in the army and navy also attended, but not as voting members. The council provided a forum for the discussion of "all questions relating to national defense."[15] Its members often assembled prior to a meeting of the Council of Ministers. Though both councils only provided advice, they nonetheless played large roles in shaping national security policy and influencing military strategy and planning. The Superior Council of War played a particularly important role in giving voice to the army's concerns and providing a forum in which senior officers could provide military advice to the government.

By 1910, the French had improved the efficiency of their high command, but political leaders still refused to appoint a general officer to command the army during peacetime. Instead, they appointed one general officer as chief of staff of the army and another as the vice president of the Superior Council of War. The former worked closely with the minister of war and administered the day-to-day activities of the army while the latter had no command authority but nonetheless had responsibilities associated with preparing the army for war. Known as the generalissimo, the vice-president of the Superior Council was the designated commander of French forces in the event of war. The Superior Council included generals who would command France's forces in war, but neither the council nor the vice president could make decisions. They could only provide advice to the minister of war. Although this arrangement ensured that the minister of war retained great power over the army, it maintained the fragmentation of the high command and resulted in numerous inefficiencies, particularly because the chief of staff worked independently from the vice president of the Superior Council.[16]

The French moved more quickly to improve the education and training of staff officers. Prior to 1870, the French had a "closed corps" of general staff officers who became staff specialists after brief service in the combat

[14] *J.O.* (1872), p. 5165; *J.O.* (1888), p. 1964.

[15] C.S.D.N., Historique et organisation de la défense nationale de 1906 à 1939–1940, SHD 2N1; C.S.D.N., Décret sur l'organisation du C.S.D.N., 28 juillet 1911, SHD 2N1.

[16] David B. Ralston, *The Army of the Republic: The Place of the Military in the Political Evolution of France, 1871–1914* (Cambridge, MA, 1967), pp. 190–93, 327–29, 331, 335–37; Douglas Porch, *The March to the Marne: The French Army, 1871–1914* (Cambridge, 1981), pp. 170–76.

arms and two years of study at the "school of application for the general staff." Dissatisfaction with the staff officers' performance in 1870–71, however, resulted in major changes. Beginning in 1876 the French selected highly qualified lieutenants and captains to attend the École Supérieure de Guerre and then to rotate between staff and line positions. Of the 290 officers authorized to apply in the first year of the new course, only 72 were selected. These officers, and their successors, had to study operations and logistics, as well as the German language. They also spent a great deal of time on terrain walks, devoting, for example, 56 days in 1888 to studying the eastern frontier.[17] As for the faculty, some of France's most distinguished leaders of World War I taught at the École de Guerre. Officers such as Ferdinand Foch, Philippe Pétain and Eugène Debeney kept students busy composing orders for military operations and mastering the details of logistics and movement. Though the École Supérieure de Guerre may have been, as some critics charged, obsessed with minutiae and better at producing bureaucrats than leaders, its graduates demonstrated strong skills in planning, troop movements, supply, and intelligence.

After the German gunboat *Panther* appeared off the Moroccan port of Agadir in 1911, the French moved quickly to improve the efficiency of their fragmented high command. Amidst the tense international situation, a decree on 28 July named General Joffre chief of the General Staff and reorganized the high command, placing it – in the words of Adolphe Messimy, the minister of war – "completely and without reserve under Joffre's direction."[18] Messimy appointed General Yvon Dubail chief of staff of the army and, except for issues dealing with personnel and logistical support, made him subordinate to Joffre. When Alexandre Millerand became minister of war in early 1912, he reduced the fragmentation of the high command further. He did not want the chief of staff of the army to deal with the minister of war rather than the chief of the General Staff on issues dealing with personnel and logistical support. He abolished the position of chief of staff of the army, created a deputy chief of the General Staff who worked for Joffre, and placed issues of personnel and logistics under the chief of the General Staff.

This change gave Joffre enormous powers over the army by shifting powers from the minister of war and chief of staff of the army to the

17 *Centenaire de l'école supérieure de guerre, 1876–1976* (Paris, 1976), pp. 11–17.

18 *J.O.* (29 juillet 1911), pp. 6444–6445; Adolphe Messimy, *Mes souvenirs* (Paris, 1937), p. 82.

chief of the General Staff. In his memoirs Joffre noted: "It was the first time that such powers were confided to a single man; I had authority over the training of the army, its doctrine, its regulations, its mobilization, its concentration." He added, "For the first time . . . , the leader [who would be] responsible [for the army] in wartime would have the authority in peacetime to prepare for war."[19] While the minister of war remained the legal head of the army and the acknowledged army chief, Joffre had the power and the trained staff officers to do the hard work associated with planning for war. Moreover, by choosing Joffre and giving him enormous powers, Messimy and Millerand set the French army firmly on the path of the *offensive à outrance*. When Messimy had considered candidates, the leading contenders were Generals Joffre, Joseph Gallieni, and Paul Pau. Though the three contenders did not differ substantially in 1911 in their attitudes toward the offensive, subsequent events would demonstrate Joffre's unwavering faith in that option.

Joffre turned out to be an exceptionally strong-willed officer who had no qualms about imposing his ideas on the French army. In a variety of assignments, he had demonstrated outstanding organizational skills and an unusual capacity to handle positions of great responsibility. He performed well as an engineer in the construction of fortifications on France's northeastern frontier and distinguished himself further with service in the colonies. In subsequent assignments he commanded an infantry division and a corps in France. While serving as director of Engineers and then as director of Support Services, he gained an understanding of logistics and railway transportation, an understanding that served him well in subsequent years. In January 1910, he joined the Superior Council of War and became its youngest member. As chief of the General Staff, he proved to be a master of details and a stickler for bureaucratic procedures. The many documents in the French archives that bear his initials suggest he saw every important message or study that came to his headquarters during the war, and his numerous edits suggest he ensured all of them supported his policies and decisions. He preferred receiving proposals in writing from his staff and, after receiving a proposal, would ponder strategic or operational choices carefully before he made a decision. He showed no patience for those who failed or who proved incompetent and sometimes exploded in anger when a subordinate questioned a decision or indirectly criticized him. Yet, he could not impose his ideas on every

[19] Joseph Joffre, *Mémoires du maréchal Joffre (1910–1917)*, 2 vols. (Paris, 1932), 1:28; Joseph Caillaux, *Mes mémoires*, 3 vols. (Paris, 1943), 2:209, 211–12.

aspect of the French military, for resistance from bureaus within the General Staff and inertia in the army ensured some limitations on changes. In the years between 1911 and 1914 his ideas nonetheless shaped the main outline of the army, especially in its formulation of doctrine, development of weapons, and preparation of personnel.

Intelligence and the Planning Process

In the years following 1871, the French General Staff devoted a great deal of time to planning and preparing for war and completed seventeen formal plans, as well as several variations of those plans. Early plans, which were concentration plans and not war plans, emphasized the defense and took advantage of the Meuse River and the branches of the Moselle River which ran almost parallel to the northeastern border. When the Germans took Alsace and Lorraine, the French lost the regions where they had long expected to defend France. To fortify the new frontier, they completed fortresses between Verdun and Belfort in the late 1880s, and to hasten the movement of units to the frontier they improved railways running from Paris to northeastern France. These improvements enabled French leaders to consider attacking after defending initially, but not until Plan XI was completed in August 1891 did they anticipate having a choice between attacking or defending at the beginning of a war with Germany. After the completion of the Franco-Russian alliance in the early 1890s, the French began to consider launching an offensive into Germany. Plans XI through XVI, nonetheless, rested primarily on a defensive-offensive strategy in which French forces initially defended and then counterattacked.[20]

Beginning in 1888, the French paid more attention to the possibility of the Germans moving north of Verdun or advancing through Belgium, but they initially had nothing more than covering forces north of Verdun or along the Belgian frontier. As early as February 1892, Plan XII included a contingency for the Germans violating Belgian neutrality. Subsequent plans also took into account the possibility of the Germans passing through Belgium to the North of Verdun. With each new plan the French increased the number of forces north and northwest of Verdun. In 1908 they foresaw the Germans sending two armies around the flank of French fortifications and passing through Luxembourg and eastern Belgium, one army emerging from the Ardennes at Verdun and the other

[20] *Les armées françaises dans la grande guerre*, 11 "*tomes*," 111 annexes (Paris: Imprimerie nationale, 1922–1937) 1/1, pp. 6–7.

at Sedan.[21] Because they did not know for sure where the Germans might strike or how deep they might advance through Belgium, they decided not to spread their forces along the frontier of eastern Belgium and chose instead to create a powerful force for a counterattack. They organized a new army, Sixth Army, and placed it near Châlons, eighty kilometers west of Verdun, so it could move easily toward Toul-Épinal on the right, Verdun on the left, or Sedan-Mézières farther on the far left. The new Plan XVI was completed in March 1909.

Over the years, the French high command received intelligence suggesting Germany had great interest in Belgium. Some of the intelligence came directly or indirectly from sources in Germany, but some also came from agents obtaining important documents, such as the German mobilization plans of 1907 and 1913 or copies of different war games conducted by the Germans in 1912 and 1913.[22] Although mobilization plans and war games did not provide explicit information about German capabilities and intentions, they provided important insights into German thinking. French intelligence also acquired a copy of the 1911 handbook for German general staff officers. Unlike the 1902 edition, it had information about the military forces of Great Britain, Belgium, and Holland.[23] To gain insights into German capabilities, the French carefully analyzed the Germans' strengthening fortifications in the Metz-Thionville region and improving their railway system. Analysts concluded that the Germans favored an advance through Belgium and were improving fortifications around Metz-Thionville to protect their center. Such an advance, they believed, would be part of a German campaign to defeat France quickly and then turn toward Russia. In May 1914 analysts, using intelligence provided by the Russians, completed an extensive study on German strategic and operational alternatives in eastern Europe. The officers concluded that the Germans would concentrate on "crushing" the French first and then turn toward Russia.[24]

A serious discussion about the Germans moving through Belgium occurred when General Victor Michel became vice-president of the

[21] Note lue par le général de Lacroix, 15 février 1908, *AFGG*, 11–2, pp. 6–7.

[22] See: Kriegspiel du XVIIIe corps d'armée, 17 janvier 1913, SHD 7N436 (Supplement). In addition to information about seven different war games in 1912 and 1913, this carton also has information about war games in 1896, 1899, 1901, 1902, 1904, and 1911. Other assessments are in SHD 7N1538.

[23] Commentaires du document S.R. n1 1685 du 20 Avril 1911 (Aide mémoire de l'officier d'état-major en Allemagne), n.d., SHD 7N436 (Supplement), pp. 3, 16.

[24] Étude relative au théatre d'opérations russo-allemandes, mai 1914, SHD 7N1538.

Superior Council of War in 1910. He foresaw the Germans seeking a victory in central Belgium, not in Lorraine or eastern Belgium. To respond to this threat, he suggested a new strategy: defend on the right from Belfort to Mézières and launch a "vigorous offensive" on the left toward Antwerp, Brussels, and Namur.[25] Since French forces could not occupy such a vast front without the full and complete integration of the reserves, Michel also proposed a reorganization of the army and a mixing of active and reserve forces. He presented his ideas on reorganizing the army to the Superior Council of War on 19 July 1911, but his ideas were overwhelmingly rejected. The army's senior generals favored the existing organization and believed that "improvised" units would be incapable of immediate offensive operations.[26] Minister of War Messimy, who had strong reservations about the proposal, quickly dismissed Michel and replaced him with Joffre.

On 6 September, only six weeks after becoming chief of the General Staff, Joffre published a new variation to Plan XVI. Dismissing Michel's concept for defending along the entire Belgian frontier, Joffre moved the center of mass of French forces farther north and spread them so they could cover more, but not the entire frontier. Drawing a line between Paris and Metz, he shifted four active corps from south of the line to the North and three reserve divisions and one cavalry division from north of the line to the South. He then pushed his reserves closer to the Belgian and Luxembourg frontiers. These changes placed more forces near the Belgian border and provided additional protection against the possibility of the Germans attempting to envelop the French flank by driving around Verdun or as deep as Sedan or Mézières.[27]

During the next several months, Joffre continued to review French strategy and to contemplate the possibility of the Germans advancing through Belgium. The French staff focused on several alternatives for the Germans and eventually decided they could attack anywhere along the northeastern frontier of France, but would more likely defend in Lorraine and attack through Belgium. At a meeting of the Superior Council of National Defense on 9 January 1912 Joffre raised the question of French forces advancing into Belgium before the Germans, but the

[25] C.S.G., Concentration et plan d'opérations, février 1911, *AFGG*, 11–3, pp. 7–11.

[26] P.V., C.S.G., Séance du mercredi, 19 juillet 1911, SHD 1N10; Extraits du P.V., C.S.G., Séance du mercredi, 19 juillet 1911 Fonds Messimy, AN 509 AP/5.

[27] Répartition schématique des forces pendant la concentration sous le plan XVI et le variant n1 1, n.d.; Le concentration d'après le plan XVI (1 mars 1909) et la variant n1 (Sept. 1911), n.d. Both are in Fonds Messimy, AN 509 AP/5.

Council refused to permit a violation of Belgian neutrality. With Joffre's concurrence, the Council agreed that French forces could "penetrate the territory of Belgium at the first news of the violation of that territory by the Germany army."[28] The Council also considered France's "secondary frontiers," the Alps and Pyrenees, and agreed that reserve or territorial forces could provide sufficient protection there. Focusing primarily on the prospects of a war with Germany, the Council had few comments about naval forces except for their importance in transporting troops from North Africa.[29]

As the French analyzed the implications of a German advance through Belgium, they carefully examined German efforts to improve the quality of their reserves and thereby enable them to advance alongside active units. One important source was Germany's mobilization plan of October 1913, a copy of which was obtained by French agents. The mobilization plan included the phrase, "[R]eserve troops will be employed the same as active troops."[30] French intelligence officers, however, were not impressed with the quality of German reserve units; they believed the reserve units needed additional "capable" officers, had insufficient artillery, and could not deliver the "shock" of front-line troops. They expected the Germans to place their reserves in the second line and not to integrate their active and reserve units. Given other operational demands such as defending against the Russians, the Germans did not have sufficient strength or active corps, said the intelligence officers, to extend their drive across most of Belgium. The Germans could drive deep into Belgium only if they weakened their center in eastern Belgium and Luxembourg.[31]

As Joffre considered an attack into Belgium against what he thought would be the weakened German center, he decided not to inform his political superiors or the army commanders who would conduct the operation. He wanted no interference from political leaders. By providing operational alternatives in a concentration plan (Plan XVII), he believed he could give his subordinates sufficient guidance for them to plan and prepare. He noted in his memoirs,

> I preferred to say nothing [about an advance into Belgium] in an operations plan, contenting myself with a concentration [plan] with various alternatives. And I

[28] C.S.D.N., Séance du 9 janvier 1912, pp. 4–5, AN 509AP/5. See Guy Pédroncini, "Stratégie et relations internationales: La séance du 9 janvier 1912 au conseil supérieur de la défense nationale," *Revue d'histoire diplomatique* 91 (1977), pp. 143–58.

[29] C.S.D.N., Séance du 9 janvier 1912, AN 509AP/5; Caillaux, *Mes mémoires*, 2:213–14.

[30] Joffre, *Mémoires*, 1:249.

[31] Ibid., 1:135, 139, 278; *AFGG*, 1/1, p. 39.

confined myself to announcing my intention to attack in the general direction of the northeast as soon as all French forces were assembled.[32]

In April 1913, the Superior Council of War approved Joffre's "elaboration" of a new concentration plan, and in February 1914 his staff finished the main parts of Plan XVII and issued copies to each army commander. The staff finally completed all annexes in the plan on 1 May.[33] The new plan envisaged France's concentrating five armies in the Northeast and clearly stated, "The intention of the commander-in-chief is to deliver, with all forces assembled, an attack against the German armies."[34] Other than this bold statement, the plan said nothing about Joffre's intentions and included only instructions for concentrating forces in the Northeast and options for attacks. From right to left, Joffre placed First, Second, Third, and Fifth armies along the German, Luxembourg, and Belgian frontiers. Fourth Army remained in reserve behind Second and Third armies. Fifth Army occupied a broad front stretching across the Luxembourg-Belgian frontier and extending west beyond Mézières to Hirson. Joffre also had four "Reserve Division Groups," each with three divisions. After positioning the First Group on the right of First Army and the Fourth Group on the left of Fifth Army, he placed the Second and Third groups behind Second and Third armies. Plan XVII thus had about the same forces along the Luxembourg and Belgian frontiers as had Joffre's modified version of Plan XVI.

Under the new plan, French forces occupied a central position from which Joffre could launch an offensive toward his left, center or right. Each army commander received specific instructions in Plan XVII which included the mission of the armies on his flanks and provided options for possible operations. On the right, First and Second armies prepared attacks south of Metz-Thionville. While First Army focused on driving toward Sarrebourg and then Sarreguemines, Second Army focused on Saarbrücken. In the center Third Army focused on Metz-Thionville, and on the left, Fifth Army prepared to advance into Luxembourg or Belgium. The direction of Fifth Army's move depended on whether the Germans entered neutral Luxembourg and Belgium. If the Germans did not violate the territory of Luxembourg or Belgium, it would attack north of

32 Joffre, *Mémoires*, 1:190.
33 P.V., C.S.G., Séance du vendredi, 18 avril 1913, p. 66, SHD 1N10; Joffre, *Mémoires*, 1:169–80, 188–89. An original copy of the plan, including signatures of all the major commanders, is located in SHD 7N1778.
34 E.M.A., Plan XVII, Directives pour la concentration, 7 février 1914, *AFGG* 11–8, pp. 21–22.

Thionville into Luxembourg but would retain sufficient forces along the Belgian frontier to protect against a subsequent German drive deep into Belgium. If the Germans entered Belgium, Fifth Army would advance north into Belgium toward Florenville and then Neufchâteau. In this latter case, Fourth Army would enter the line between Fifth and Third armies and march into Belgium toward Arlon. If the Germans did not violate the territory of Luxembourg or Belgium, Fourth Army would enter the line between Third and Second armies and participate in the Lorraine attack. Even if the army commanders did not know Joffre's overall strategy, they knew enough about the operational alternatives to complete their planning.

A secret annex to Plan XVII dealt with the British.[35] Since the British arrival could be delayed by the timing of their decision to mobilize or by problems in transportation, Joffre knew the British might not participate in initial operations. He hoped the British would arrive in time to take part in the campaign, but he did not allow their presence or absence to affect his strategy or his operational planning. He asked them to occupy a position west of Mézières on the left of Fifth Army, where they could extend the French line or, if needed, assist Fifth Army.[36]

Other Preparations Associated with Planning

Joffre recognized that success against the Germans depended on more than a concentration plan. He knew the French army had to have operational and tactical units that could support an offensive strategy fulfilling the requirements of the Franco-Russian alliance. Above all, he believed those units had to be capable of rapid offensive operations in a short war. Beginning what he later called the "transformation" of the French army, he turned his attention to formulating a viable offensive doctrine, developing suitable weapons and units, and ensuring France had highly trained officers and soldiers.[37] He knew the plan could not be developed separately from the means to execute it. As Joffre transformed the French army, he prepared it for a short war in which the army consumed more resources than in 1870–71 but far fewer than they would actually consume in 1914–18. He prepared the army for a short war even though he

[35] W, Prévisions de l'état-major de l'armée (4e bureau) relative à la durée de la concentration W, *AFGG* 11–7, p. 20.
[36] Joffre, *Mémoires*, 1:148–49.
[37] Ibid., 1:29.

acknowledged in a meeting of the Superior Council of National Defense on 21 February 1912 that a war with Germany could last for an "indefinite period." Whether the Germans or French won the initial battles, he acknowledged, it still would take many months for the initial winner to destroy his opponent's resistance completely.[38]

Despite this awareness, he made almost no preparations for a long war or for mobilizing the industrial power of France. Logisticians assumed munitions and equipment manufactured in peacetime would meet the army's needs and goods manufactured in wartime would replace those consumed. The army planned on living off its stockpile of ammunition until the daily manufacture of 75-mm rounds reached about 13,600 rounds per day, a number significantly lower than the 50,000 Joffre would request seven weeks after the war began. After the announcement of mobilization, Paris began accumulating food supplies, but this stockpiling came less from preparations for a long war than from fears of another siege such as the one of 1870–71.[39] The failure to understand the relationship between industrial production and military operations, and the unwillingness to think of anything but a short war, contributed to Joffre's paying no attention to defending France's natural resources in the first month of the war, the loss of which deprived France of 83 percent of its production of iron ore, 62 percent of its cast iron, and 60 percent of its steel.[40] In the end, the failure to prepare, or even think through, industrial mobilization forced French soldiers to fight without adequate weapons and supplies and cost thousands of them their lives.

Instead of worrying about logistics, French officers spent a great deal of time thinking about operations. After much hard work, a "logical and sensible doctrine of the offensive," to use Joffre's words, was produced.[41] The ideas of the *offensive à outrance* permeated the doctrine. A new regulation on the operations of large units (corps, army, and army group) was published in October 1913, and one on smaller units (regiment, brigade, and division) in December 1913. As for the significance of the two regulations, the commission that wrote the October 1913 version explained,

38 Ibid., 1:123n–24n.

39 Henri Sellier, A. Bruggeman, and Marcel Poete, *Paris pendant la guerre* (Paris, n.d.), p. 20.

40 France, Assemblée Nationale, Chambre des Députés, Onzième legislature, Session de 1919, vol. 71, no. 6206 (annexe), *Rapport fait au nom de la commission d'enquête (Question de Briey*, 1re partie: *Concentration de la métallurgue française sur la frontière de l'est*)(Paris, 1919), pp. 4–5.

41 Joffre, *Mémoires*, 1:37.

"The French army, returning to its traditions, accepts no law in the conduct of operations other than the offensive." The regulations emphasized, "Only the offensive yields positive results." The will-to-fight became all-important:

> Battles are above all moral contests. Defeat is inevitable when hope for victory ceases. Success will come, not to the one who has suffered the least losses, but to the one whose will is the steadiest and whose morale is the most highly tempered.[42]

The December regulations added, "Once begun, combat is pushed to the end; success depends more on the vigor and the tenacity of execution than on the skill of combined actions. All units thus are employed with the most extreme energy."[43]

The regulations also emphasized a new relationship between artillery and infantry. The report with the December 1913 edition said, "The artillery does not prepare attacks; it supports them."[44] In other words, instead of blasting enemy positions for long periods prior to an infantry assault, artillery fired primarily during an attack. Whereas previous regulations had envisaged the artillery firing many rounds and shelling enemy positions for long periods prior to an assault, the new regulations warned against the massive use of artillery and called for support only during the infantry's advance. The regulations asserted that artillery had only a limited effect against an entrenched enemy. "To force an adversary out of his cover, it is necessary to attack with the infantry."[45] The infantry regulations of April 1914 said that the "supreme weapon" of the infantry was the bayonet. After the infantrymen fixed bayonets, they would advance with their officers leading in the front and with drums and bugles sounding the charge.[46] The attacking troops would supposedly gain a superiority of fire with the rapid and intense fire of the 75-mm cannon and with a hail of bullets from their rifles. When they closed with the enemy, they would finish the fight with the bayonet. Such ideas, which were questioned by almost no one, would cost the French thousands of casualties in the opening battles of 1914.

[42] France, M.G., *Décret du 28 octobre 1913 portant règlement sur la conduite des grandes unités* (Paris, 1913), pp. 48, 7.

[43] France, M.G., *Décret du 2 décembre 1913 portant règlement sur le service des armées en campagne* (Paris, 1913), pp. 76–7.

[44] *Règlement sur le service des armées en campagne (2 décembre 1913)*, p. 15.

[45] Ibid., p. 78.

[46] France, M.G., *Règlement de manoeuvre d'infanterie du 20 avril 1914* (Paris, 1914), pp. 69, 139–40.

As the French developed new doctrine, they recognized the requirement for heavy artillery. German improvement of their 77-mm guns and their adoption of 105-mm and 150-mm guns made it hard to deny the need for longer-range pieces in the French arsenal.[47] In July 1911 the Superior Council of War supported the adoption of a "light howitzer" that would have a higher trajectory than that of the 75-mm gun and could "accompany" the infantry in attacks.[48] Although the French had developed a 155-mm short-range cannon, the generals wanted a more mobile 120-mm howitzer, capable of following the infantry closely. In essence, they foresaw a war of movement and wanted mobile heavy artillery that could keep up with the infantry in the field.

In October 1911, a special commission formed by the minister of war also recommended a light howitzer with a round larger than the 75-mm and a range of twelve to thirteen kilometers. Though French artillerists recognized the requirement for heavy artillery, they did not want to give up their legendary 75-mm rapid-firing artillery piece. They had great confidence in the light piece and knew it could keep up with the highly mobile French infantry. The technical services also objected to introducing more calibers and thereby multiplying supply problems. When critics emphasized the vulnerability of 75-mm batteries to enemy counter-battery fire, artillery officers argued that 75-mm batteries could displace rapidly and thereby protect themselves against the longer-range fire of the German pieces. When critics of the 75-mm gun noted that its flat trajectory kept it from occupying and firing from defilade positions and prevented its rounds from hitting targets on the reverse slope of hills, technicians came up with a modification to the 75-mm round (*plaquette Maladrin*) that gave the round a more curved trajectory and supposedly enabled it to strike defilade targets. Officers on the General Staff described this as a "simple and ingenious" modification.[49] After testing and modifying pieces already being manufactured in France for the Bulgarians and Russians, the French ordered 220 105-mm howitzers that had a range of 12.3 kilometers. Objections from technicians, however, convinced the minister of war to reduce the order from 220 to 36, all of which appeared just as the war started. The French also initiated a program to modify existing 120-mm and 155-mm long-range guns and provide them greater

47 P.V., C.S.G., Séance du mercredi, 9 juin 1909, pp. 90–3, SHD 1N10; Extraits du P.V., C.S.G., Séance du mercredi, 19 juillet 1911, Fonds Messimy, AN 509 AP/5.

48 P.V., C.S.G., Séance du mercredi, 19 juillet 1911, p. 197, SHD 1N10.

49 État-Major de l'Armée, 3e Bureau, Plan XVII, Bases du Plan, 2 mai 1913, SHD 1N11.

mobility and range. Some improvements were made, but they occurred too late to affect the war's first battles.

Joffre later insisted he had worked diligently to obtain heavy artillery for the French army, but at the July 1911 meeting of the Superior Council of War which supported the adoption of a "light howitzer," he had argued for an artillery piece that could "accompany" the infantry in attacks. Similarly, at a meeting of the Superior Council of War in October 1913 he opposed placing heavy artillery in corps and instead favored forming heavy artillery regiments and placing them at field-army level.[50] Given the reluctance of the artillery community to demand heavier, longer-range pieces and given its success in stymieing reform, he escapes some of the blame for the artillery's weakness, but he never used his energy and his authority to improve artillery the same way he used them to change doctrine. He, like most French officers, could foresee situations in which heavy artillery would be useful, but since he expected a series of highly mobile battles, he did not want to burden corps or divisions with relatively immobile heavy artillery. The French had complete confidence in their rapid-firing 75-mm cannon, the best in the world, and few officers questioned having 120 75-mm cannon in a corps while the Germans had 108 77-mm, 36 105-mm, and 16 150-mm.

In addition to devoting considerable effort to doctrinal and material questions, the French spent an enormous amount of time thinking through the mobilization and training of personnel. Their primary concerns were avoiding the chaos of 1870 and having well-trained soldiers ready for offensive operations in the opening battles of a war. The acrimonious debate on the eve of the war over increasing the term of conscripts' service from two to three years originated not in the army's desire to put more men in uniform during a war, but with its desire to have as many men under arms in the opening battles as Germany and to have these men ready for operations immediately. The army's leaders considered conscripts with only two years' service inadequately trained, and they believed the departure each year of a class of conscripts left the army in a "precarious situation" with too many inexperienced soldiers and non-commissioned officers.[51] Though the 1905 law had been accompanied by promises to increase the number of non-commissioned

[50] P.V., C.S.G., Séance du mercredi, 15 octobre 1913, pp. 4, 7, SHD 1N11.

[51] Joffre, *Mémoires*, 1:86. On the subject of three years of service, see Gerd Krumeich, *Armaments and Politics in France on the Eve of the First World War* (Dover, NH, 1984).

officers, those increases never appeared. As a result, the army suffered from a critical shortage of small-unit leaders; many of its corporals and sergeants were only in their second year of service. Under such circumstances, covering the frontier in an emergency or launching offensive operations immediately, the army's leaders reasoned, would be extremely difficult.

The army's complaints fell on deaf ears until the Germans passed laws in 1912 and 1913 expanding the size of their active army. In a meeting of the Superior Council of National Defense in March 1913, Joffre explained that Germany's reforms had accelerated the pace of its mobilization and would enable its forces to "push aside" France's covering force and disrupt the mobilization of the remainder of the army.[52] The obvious disparity between German and French active forces at a time of increasing international tension resulted in quick political action. Despite fierce opposition, the Chamber of Deputies in July approved changing the term of service from two years to three years, and the Senate passed the same measure in August. According to Joffre's figures, the law increased the number of immediately available soldiers in the French army to 700,000 while the Germans had 870,000. With about 175,000 German soldiers remaining in the East to face the Russians, France and Germany would have, he believed, approximately the same number of soldiers in the West in the opening battles of a war.[53] If the British arrived in time, French and British forces could outnumber the Germans in a campaign in western Europe. Yet, the change in term of service did not increase the total number of soldiers in French uniforms after mobilization. It affected only the number immediately available and their level of training at the start of a war.

Other reforms contributed to the transformation of the French army. Serious efforts to add aviation assets and determine the new weapon's capabilities and limitations clearly demonstrate the army's interest in something other than bayonet charges. In an era of enormous technological change the French had 43 pilots in November 1910, plus another 96 capable of flying aircraft. By December 1911, the French had 261 pilots. A special study of aircraft in January 1912 recommended adding 328 aircraft to the 120 France had in its inventory and the 40 already on order. The report concluded that aircraft in the future would be

[52] P.V., C.S.G., Séance du 4 mars 1913, SHD 1N11.

[53] Joffre, *Mémoires*, 1:95; Raymond Poincaré, *Au service de la France*, vol. 3, *L'Europe sous les armes, 1913* (Paris, 1926), p. 215.

"an indispensable instrument for our armies in the field."[54] When the Superior Council of War met, it supported ordering the additional aircraft, and upon mobilization in August 1914, the French had about 140 airplanes organized into squadrons of five or six planes.[55]

They also added training areas. The parade grounds common to many military installations proved completely inadequate for tactical training at the beginning of the twentieth century. The longer-range rifles and artillery that were added to the French arsenal in the decades before 1914 could not be fired on small training areas, and the requirement for immediate offensive operations required larger areas in which regiments and divisions could train and maneuver. The Superior Council of National Defense observed in January 1912: "[T]roops need large areas designed especially for them and capable of supporting combined operations of units from all arms."[56] Adopting three years of military service added to the number of troops, including reservists, needing to visit training areas each year, but the high cost of additional land spread the expansion of training areas over a longer term. In his memoirs Joffre stressed the importance of the larger training areas and the difficulty of acquiring them, but he also noted that some improvements had occurred by the eve of the war.[57]

The French used maneuvers to assess and improve their readiness. While minister of war in 1888–91, Freycinet had renewed emphasis on their importance and, for example, had 100,000 troops participate in maneuvers in 1891. In subsequent years autumn maneuvers remained a fixture in the French army, for they gave commanders an opportunity to move their units out of barracks and practice, as well as test, important operational and logistical methods.[58] The French conducted different types of maneuvers, including force-on-force, "cadre," and map exercises. In 1912, Joffre used the autumn maneuvers to test aviation capabilities and gained important insights into the use of air for strategic, tactical, and

[54] Section Permanente de Aéronautique Militaire, Rapport du Général Roques, 27 janvier 1912, SHD 1N17.

[55] P.V., C.S.G., Séance du 25 janvier 1912, SHD 1N10; M.G., É.-M. de l'Armée de Terre, Pierre Guinard, Jean-Claude Devos, and Jean Nicot, *Inventaire Sommaire des Archives de la Guerre* (Troyes, 1975), p. 174.

[56] P.V., C.S.D.N., Séance du 9 janvier 1912, Fonds Messimy, AN 509 AP/5.

[57] Joffre, *Mémoires*, 1:79–84.

[58] On the background and purpose of maneuvers, see G. G.(Georges Gilbert), "Manoeuvres d'autumne," *Nouvelle Revue*, 1 August 1890, pp. 449–73; Exercices Spéciaux pour le Haut Commandement et les états-majors, 8 mai 1914, SHD 7N1930.

"special" reconnaissance.[59] He also put four infantry corps, two cavalry divisions and a reserve division, as well as appropriate supply and service units, through their paces. He divided the forces into two "armies" and had them maneuver against each other. The maneuvers raised serious questions about the capability of reserve units and the readiness of the French army.[60] In his memoirs Joffre praised the "offensive spirit" of the soldiers, but he noted "grave errors" in "uncoordinated actions, imprudent maneuvers. . . . , [and] the employment of artillery." Maneuvers in the previous year, Joffre said, had revealed many of the same errors as well as "grave insufficiencies in command." He explained, "Many of our generals proved themselves to be incapable of adapting to the conditions of modern war. . . ."[61] He attempted to improve the army's performance by having units spend more time in training camps and by conducting in July 1914, amidst the Balkan crisis, a large command post exercise that included cadre from two armies and four corps.[62] But the war started before he was able to make much-needed improvements. His relief of two army, ten corps, and thirty-eight division commanders in the first month of fighting was undoubtedly influenced by the negative opinions he formed during the maneuvers of 1912 and 1913.[63]

Despite evidence of "grave insufficiencies," Joffre had great confidence in the ability of his forces to conduct offensive operations shortly after mobilization. As French political leaders took the final steps in adopting three years of military service, he participated in staff talks with the Russians and assured them of the reliability and capability of the French army. He promised that France would engage "nearly all its forces" on the northeastern frontier and would commence offensive operations on the eleventh day after mobilization.[64] His promise to start offensive operations so soon after mobilization was undoubtedly linked to the increase in the term of military service.

59 Le Chef d'État-Major de l'Armée, Extraits du Rapport sur l'Aviation à l'armée de l'ouest pendant les manoeuvres de 1912, n.d, SHD 1N17.

60 P.V., C.S.G., Séance du vendredi, 18 avril 1913, p. 63, SHD 1N10.

61 Joffre, *Mémoires*, 1:37–8.

62 Voyage d'Armée et d'État-Major de l'Armée et du Centre des Hautes Études Militaires, Instruction général sur l'organisation et fonctionnement du Voyage d'Armée avec Cadres, n.d, SHD 7N1930.

63 Joffre, *Mémoires*, 1:421; Pierre Rocolle, *L'Hécatombe des généraux* (Paris, 1980), p. 262.

64 P.V. des entretiens du mois d'août 1913, 24 août 1913, *D.D.F.*, 3e série, vol. 3, annex no. 79, pp. 88, 90.

Much depended, however, on the ability of the Russians to deliver a powerful offensive at the same time as the French. A note written in August 1912 informed the premier:

> The military value of the Alliance resides above all in the possibility of obtaining simultaneous attacks from the east and west. But, this ideal is far from being realized: the Russian armies are actually very far behind the French armies.[65]

With considerable assistance and advice from the French, the Russians made improvements over the next two years, especially in their development of heavy artillery and their construction of railways. A study done in the preparation of Plan XVII noted that the Russians were getting better "day by day."[66]

Frequent reports from France's military attaché in St. Petersburg provided more information about improvements in the Russian army, and Joffre's visit to Russia for three weeks in August 1913 gave him a chance to view Russian training and maneuvers and to appraise the Russians' readiness for an immediate offensive. He clearly was not impressed and in his memoirs noted that the maneuvers were conducted like a parade "without attempting to take into account the realities of war."[67] Attending the maneuvers gave him an opportunity to suggest modifications. He brought with him a long memorandum written by his staff suggesting changes in Russian strategy and operations. The following May, only three months before the outbreak of the war, the Russians conducted a war game at Kiev to test Joffre's new ideas, but it was too late to make significant changes.[68] Joffre's reservations about the capability of the Russians were nonetheless not significant enough for him to abandon or modify the strategy of simultaneous offensives by the Russians and French. He had watched the Russians get better "day by day," and his comments in several staff talks with them suggest that he thought that the Russians and French were strong enough in 1914 to prevail against the Germans and Austrians.

[65] Note sur l'action militaire de la Russie, n.d., SHD 7N1538.

[66] État-Major de l'Armée, Plan XVII: Bases du Plan, 2 mai 1913, p. 5, SHD 1N11.

[67] Joffre, *Mémoires*, 1:132–33.

[68] Étude relative à l'importance à la repartition et à l'emploi des forces allemandes sur le théatre d'opérations Russo-Allemand, mai 1914, SHD 7N1535; Louis Garros, "En marge de l'alliance Franco-Russe, 1902–1914," *Revue historique de l'armée* 6 (avril-juin 1950), pp. 42–4.

The Campaign

In July–August 1914, political and military leaders thought France was well prepared for a war. Some of them, such as War Minister Messimy, welcomed the conflict because it gave France an opportunity to regain Alsace and Lorraine and exact revenge for the defeat of 1870–71. When Messimy met with Joffre on the evening of 24 July and told the general that France might go to war, Joffre responded calmly, "Very well, sir, we will do it if it is necessary." Messimy energetically shook his hand and said, "Bravo!"[69] This enthusiasm contrasted sharply with the restraint of August 1911 when Joffre had told Premier Joseph Caillaux that France did not have a 70 percent chance of victory.[70] In the intervening three years France had made many improvements in its army and had devoted great attention to planning for a war against Germany. That planning rested above all upon the strategic concept of France and Russia attacking Germany simultaneously.

During the July Crisis, France moved cautiously toward full mobilization. Incremental steps included recalling all officers from leave, alerting naval forces to remain in port and fill their ammunition stocks, warning troops in Algeria and Morocco to prepare for movement to France, and checking wireless communications with the Russians. After Austria-Hungary declared war on Serbia on 28 July, France accelerated its preparations, but to avoid openly provocative steps the Council of Ministers denied Joffre's request to send covering forces to the border. On 30 July the Council approved placing part of France's covering forces along the border but insisted they be kept ten kilometers from the frontier. France would mobilize, as Abel Ferry wrote, but "not concentrate." The Council ruled out summoning reservists and allowed movement only of those troops close enough to march, rather than ride the railway, to the frontier. Joffre considered the actions far too timid, but the careful steps reinforced the appearance of France as a victim.

As Austria-Hungary and Russia began complete mobilization on 30–31 July, the Council of Ministers approved sending the entire covering force to the border, but again ordered it to remain ten kilometers from the frontier. Poincaré explained the continued caution, "It is better to have war declared on us."[71] The Russian military attaché in Paris, however,

[69] Joffre, *Mémoires*, 1:207–08.
[70] Ibid., 1:15–16.
[71] Poincaré, Notes journalières, 1 août 1914, p. 138, BNF NAF 16027.

wired St. Petersburg that Messimy had told him "enthusiastically" about the "firm decision" of the government to go to war and asked him to remind Russian authorities that the French high command wanted the Russians to direct "all" their efforts against Germany.[72] On 1 August, after a meeting between the premier of France and the German ambassador, the Council of Ministers authorized the minister of war to issue the mobilization order that afternoon. This expanded the army from 884,000 soldiers to 2,689,000.[73] Unlike the chaotic mobilization of 1870, everything went remarkably smoothly as soldiers reported to their units, and staff officers requisitioned from civilians a wide variety of material ranging from horses to dirigibles.

As French units began concentrating on the northeastern frontier in accordance with Plan XVII, Germany declared war. Joffre and Messimy agreed that while the government determined the political objectives of the war, its conduct belonged "exclusively" to the general-in-chief. This meant strategy was completely in Joffre's hands, but as he acknowledges in his memoirs, "It was still too early to announce formally my intention to operate in Belgium."[74] He could not complete his strategy until after the enemy began advancing and until he was sure the Russians would launch an offensive into Germany. Grand Duke Nicholas, who was named commander of Russian forces on 2 August, soon dispelled fears of Russian timidity: "I am resolved to launch an offensive as soon as possible, and I will make an all-out attack."[75] After subsequent messages from the French ambassador to Russia indicated that the grand duke would begin his offensive on 14 August, Joffre resolved to launch an attack into Lorraine on the same day. To reassure the French people, he began the campaign, however, with a raid into Alsace on 7 August. Two infantry divisions and a brigade managed to reach Mulhouse, but were quickly driven back.

On 8 August, Joffre issued his General Instructions No. 1 and finally revealed his strategy, the goal of which was destruction of enemy forces, not occupation of territory. In essence, he would jab with his right and attempt a knockout blow with his left. While the offensive on the right fixed German forces in Alsace and Lorraine and drew enemy forces to

[72] Télégramme secret de l'ambassadeur à Paris, 18/31 juillet 1914, in René Marchand, *Un livre noir*, 2 vols. (Paris, 1922) 2:294; Messimy, *Mes Souvenirs*, pp. 181–87.

[73] Guinard, Devos, and Nicot, *Inventaire Sommaire des Archives de la Guerre*, p. 206.

[74] Joffre, *Mémoires*, 1:236.

[75] Ambassadeur de France en Russie, Offensive de l'armée russe, 5 août 1914, *AFGG*, 11–52, p. 85.

the South, the main attack on his left would strike the German center. The main attack, Joffre believed, would avoid the powerful enemy force driving toward his left through central Belgium and strike the enemy's less dense, more vulnerable center in eastern Belgium. After defeating the enemy forces in this area and unhinging those advancing into central Belgium, he wanted one of the armies driving into eastern Belgium to turn west and strike the enemy's main force in the left flank and rear. On his right he sent the First and Second armies into Lorraine, south of the German fortifications of Metz-Thionville, and on his left he sent the Third and Fourth armies into eastern Belgium, north of the German fortifications of Metz-Thionville. On the left of Third and Fourth armies, he placed Fifth Army in the vicinity of Mézières and aimed it toward the powerful enemy force driving through central Belgium. Aware of the possibility of a deep envelopment by the Germans, he also ordered the Fourth Reserve Division Group to occupy a "fortified position" on his extreme left and guard against an attack from the North or East.[76]

On the morning of the 14th, Joffre's offensive on his right began. First and Second armies moved forward easily at first, but as they advanced their units tended to disperse across an increasingly larger front. Then, on the 20th the Germans counterattacked. Because little liaison or coordination existed between First and Second armies, the French fought two separate and distinct battles, both of which they lost. As shaken elements of First and Second armies fell back in disorder, they fought several hard battles against the advancing Germans. By the evening of the 23rd, Second Army's right was twenty-five kilometers behind where it had started on the 14th. Beginning on 24 August, the French counterattacked and slowly pushed the Germans back. By early September, they occupied the approximate position from which they had begun their offensive on 14 August.

As the battle unfolded on Joffre's right, he aimed Third and Fourth armies at the enemy forces in eastern Belgium which were expected to head toward Sedan and Montmédy. By attacking and defeating this group, which he thought was smaller than the group heading toward central Belgium, he hoped to halt the advance of the German main attack. After issuing the attack order on 18 August, Joffre held Third and Fourth armies in place for the next two days. Meanwhile, he received reports

[76] G.Q.G., Instruction générale N 1, 8 août 1914, *AFGG*, 11–103, pp. 124–26; G.Q.G., Ordre particulier pour le 4e groupe de divisions de reserve, 8 août 1914, *AFGG*, 11–105, p. 127.

of "important" enemy forces crossing the Meuse just south of Liège and another group of large enemy forces advancing northwest some forty to fifty kilometers forward of the outposts of Third and Fourth armies. Still convinced that the Germans did not have enough active troops to drive deep into Belgium without weakening their center, he welcomed the reports. Finally, at 2030 hours on 20 August, the day the Germans unleashed their counterattack against First and Second armies in Lorraine, he ordered Third and Fourth armies to attack on the following day toward Arlon and Neufchâteau. To facilitate the task of Third Army, Joffre split it into two armies, Third Army and the Army of Lorraine. This allowed Third Army to concentrate on advancing into Belgium and the Army of Lorraine to guard against an enemy incursion from the vicinity of Metz.

As Third and Fourth armies charged into Belgium on 21 August, the French expected to outnumber the enemy, but this expectation proved false. The Germans had ten infantry corps in this region while the French had nine. Because the German Fourth and Fifth armies were near the center of the gigantic sweep through Belgium, they had moved more slowly than the First, Second, and Third armies which were farther to the West. When they learned of the French advance, they prepared themselves for a significant encounter. In contrast, the French knew little about the enemy to their front, had inadequate maps, and remained overly optimistic about their chances of success.

On the first day of the French advance into Belgium, Third Army encountered the Germans first, but the enemy consisted only of small detachments. On the second day of the offensive, 22 August, the French encountered more enemy forces and soon learned that the Germans had not denuded eastern Belgium of troops. Even though the enemy occupied strong positions, the French charged forward, often with little or no artillery support, and suffered heavy losses. Despite poor results on the 22nd, Joffre told the Third and Fourth commanders to resume the offensive as soon as possible on the 23rd. His entire offensive depended on the two armies making successful attacks into Belgium. But the day went very badly, and by the end of the day Third and Fourth armies had withdrawn to the approximate positions from which they had begun their offensives.

The fate of General Charles Lanrezac's Fifth Army on Joffre's far left proved no better than that of Third and Fourth armies in his center or First and Second armies on his right. Fifth Army's offensive occurred at the same time as those of Third and Fourth armies. Because Joffre did not know where the Germans would cross the Meuse, he ordered

Lanrezac to keep his forces assembled in the vicinity of Mézières and be prepared to respond to an enemy advance in the region between Namur and Mouzon. The arrival of the British added additional strength to his left. By the evening of the 20th, just prior to his launching Third and Fourth armies into eastern Belgium, Joffre was convinced he understood what the Germans were doing. With numerous reports of large German forces advancing deep into Belgium, he expected Third and Fourth armies to face relatively few enemy forces and had great hopes for their success. Even if Fifth Army and the British were outnumbered, they could delay the Germans advance, he thought, until Fourth Army delivered the coup de grace into the enemy's rear.

On 21 August, Joffre ordered Fifth Army to move forward and asked Sir John French to advance on Lanrezac's left. As the French moved forward, leading elements of German Second Army struck Lanrezac's left and elements of Third Army his right. On Lanrezac's left the British came under heavy attack at Mons from German First Army. At 2130 hours on the 23rd Lanrezac, fearing an attack in his rear, informed Joffre of his intentions to withdraw. This ended what became known as the Battle of the Frontiers and yielded the initiative completely to the Germans.

Joffre's main attack had lasted only a few days. On the right First and Second armies advanced into Lorraine on 14 August, but a German counterattack on 20 August pushed them back behind their starting positions. On the left Third and Fourth armies, paying little attention to reconnaissance or security, attacked on 21 August but withdrew to their starting positions two days later. On the far left Fifth Army moved into position on 19–20 August and began withdrawing on the 23rd. None of Joffre's attacks accomplished their strategic or operational goals. The French had miscalculated German capabilities and diffused their offensive capability across three widely separated battlefields.

In the ensuing Marne campaign, the French fought much more successfully. The light, mobile forces that proved unsuitable for brutal frontal assaults in the initial battles were ideal for moving quickly from Joffre's right to his left. Using the railways that fanned out from Paris like a spider web, Joffre started to move units shortly after Fifth Army began withdrawing. A French corps required as many as 110–120 trains and five or six days to move from his right to his left.[77] When the shifting of forces began, Joffre's initial goal was to halt the Germans moving around

[77] G.Q.G., Transports stratégiques effectués du l^er^ septembre au 19 septembre 1914, n.d., *AFGG*, 134–5297, p. 846.

his left by attacking their outer flank. Meanwhile, Fifth Army, the British Expeditionary Force, and other French forces on Joffre's left began a long withdrawal that took them initially toward and then past Paris. As French units arrived on Joffre's left, they arrived piecemeal and, as in Lorraine, often went into action with little coordination and support. When General Alexander von Kluck, commander of German First Army, made the mistake of passing to the east of Paris and opening his flank to a jumble of French units from Paris, the opportunity for the "miracle" of the Marne appeared. Though the forces attacking from Paris failed to destroy the German right, they compelled Kluck to halt his advance and pull back units to defend his right flank. This opened a hole for the timid advance of the British and the Fifth Army between German First and Second armies and forced the Germans to halt their advance and withdraw.

Although Joffre gained a remarkable victory in the "miracle of the Marne," the battles of the Frontiers and the Marne had cost the French dearly. They had lost some 329,000 soldiers, almost 20 percent of what they would lose in the entire war. The French now faced a completely unexpected situation: a long war that required enormous resources and completely different operational methods and equipment. Joffre began another transformation of the French army, one that was far more difficult than the one of 1911–14.

Conclusion

Although Joffre's strategy in August 1914 capitalized on the Franco-Russian alliance and demonstrated, in the words of Armand Fallières, president of the Third Republic, the army's resolve "to march straight for the enemy without any second thoughts," it failed miserably.[78] Joffre's limited offensive into Lorraine ensured Russian cooperation and reassured the French people, but his offensive into eastern Belgium suffered heavy losses and failed to unhinge the German sweep through central Belgium.

Prior to the launching of these offensives, clear evidence of shortcomings in the Russian and French armies did not create doubts about the wisdom of simultaneous offensives in the East and West against Germany, and intelligence reports about the Germans advancing deep into Belgium did not alter Joffre's preconceived notion of what the enemy was doing. On 23 August, however, as the failure of his strategy became evident,

[78] C.S.D.N., Séance du 9 janvier 1912, AN 509AP/5.

Joffre informed the minister of war that he had "terminated" his offensive. Refusing to acknowledge his own responsibility for the failure, he blamed others. He insisted he had placed the "main body of his army against the most sensitive point of the enemy" and had gained "numerical superiority at this point." French troops, he complained, had not demonstrated the "offensive qualities" expected of them despite "numerical superiority."[79] Later, he also argued that the transformation of the army was incomplete. He complained that the new doctrine was not accepted throughout the army, and he insisted France's lack of heavy artillery was the fault of others. Clear evidence of his dissatisfaction exists in his relief of many general officers in the first month of the war and his insisting they were not prepared for the "conditions of modern war."

Whatever Joffre's excuses may have been, he had not concentrated his army against the enemy's most vulnerable point and his soldiers had fought at a severe disadvantage. He had led the army through its transformation into a highly mobile offensive instrument, and when Messimy told him in July 1914 that France might go to war, he expressed no doubts about the army's readiness. Like numerous other commanders throughout history, he knew the army had changed considerably between 1911 and 1914 and he had to go to war with the army he had. The fact that the army was unprepared for the combat conditions it encountered in August 1914, however, had less to do with its incomplete transformation than with its being transformed into an army unsuited for the combat it initially faced.

Even though that army proved incapable of success in August 1914, it proved capable of absorbing the terrible losses of the first battles and then shifting quickly to its left for the Battle of the Marne. Yet, it also proved completely unsuited for the static, deadly battlefield it faced after the Marne. These costly failures occurred even though France had completed seventeen plans (plus several variants) between 1871 and 1914, revised its doctrine and planning, and reformed its fragmented high command. The failures also occurred even though the French had created the Superior Council of War and the Superior Council of National Defense to ensure the integration of political, economic, social, and military considerations into planning and preparation.

[79] G.Q.G., Général commandant en chef à ministre guerre, 23 août 1914, *AFGG*, 121–130, p. 112; G.Q.G., Situation générale, 23 août 1914, *AFGG*, 11–1044, p. 842; G.Q.G., Le général commandant en chef à ministre à guerre, 24 août 1914, *AFGG*, 121–149, p. 124.

Before the actual fighting, French planning produced impressive results. Even though the scope of war planning had become extraordinarily complex in an age of industrial warfare, the mobilization and organization of almost three million men in the army occurred remarkably smoothly, especially in comparison to the bumbling efforts of 1870. Mobilized units moved quickly and easily into position on the northeastern frontier and, once in position, responded effectively to the orders they received. When the bullets started flying, however, the inadequacies of French planning became obvious, and notions of bayonet charges with little artillery support evaporated quickly. The *offensive à outrance* may have accorded with French preferences and supported the strategy of simultaneous offensives, but it proved completely inappropriate in the face of modern firepower.

The lack of preparation for a long war also proved disastrous. Even though planners arranged a highly effective mobilization of personnel, they paid little or no attention to industrial mobilization or to economic requirements in something other than a short war. And in comparison to the actual demands of trench warfare, their prewar assumptions about the consumption of munitions and material appear more like the work of amateurs than the product of competent professionals. Rarely in history has concerted, dedicated planning produced such inadequate results.

In the final analysis, the failures in 1914 came less from the planning process than from the ideas that shaped the outcome. False conceptions and preconceived notions dominated French thinking. Absent Joffre and an extraordinary emphasis on the offensive and a short war, the outcome of the planning process would have been very different, but the elevation of Joffre in July 1911 ensured the dominance of the *offensive à outrance* over the army in August 1914. Planning by its very nature depends on a process, but the process is never better than the ideas and people within it. In 1914 preconceived ideas about the offensive and a short war dominated the planning process and ultimately cost the lives of thousands of French soldiers.

6

Great Britain

Keith Neilson

It is problematic, and perhaps misleading, to speak of British war plans before 1914. War plans conjure up images of well-oiled machinery automatically and inexorably committing armies to what the historian A.J.P. Taylor has called "war by timetable."[1] War plans also tend to suggest that a state has worked out provisions to take aggressive action against its foes, is obliged by alliances to take up arms in support of an ally, or, at the very least, has strategies to defend itself against its enemies. Only the last of these applies to Britain before 1914. Britain had no intention of committing aggression against anyone and was not bound by treaties, secret or otherwise, to go to war. Nor were British war plans by any means automatic responses to events. Instead, British prewar plans were contingent upon circumstances, something reflecting its peculiar position in Great Power politics.

Although the decade before 1914 is often characterized as an era when the Triple Entente of Britain, France, and Russia faced the Triple Alliance of Germany, Austria-Hungary, and Italy, this was not the case.[2] Although the Anglo-French *entente cordiale* of 1904 and the

[1] A.J.P. Taylor, *War by Timetable: How the First World War Began* (London, 1969).

[2] For such thinking, see George Monger, *The End of Isolation: British Foreign Policy 1900–1907* (London, 1963); *The Foreign Policy of Sir Edward Grey*, ed. F.H. Hinsley (Cambridge, 1977); Paul Kennedy, *The Rise of the Anglo-German Antagonism 1860–1914* (London, 1980); and John Albert White, *Transition to Global Rivalry: Alliance Diplomacy and the Quadruple Entente, 1895–1907* (Cambridge, 1995). For countering arguments, see Keith Neilson, *Britain and the Last Tsar. British Policy and Russia, 1894–1917* (Oxford, 1995); and Keith Wilson, *The Policy of the Entente: Essays on the Determinants of British Foreign Policy, 1902–14* (Cambridge, 1985).

Anglo-Russian Convention of 1907 had improved Britain's relations with its two long-standing foes, and had created in many minds (a substantial number of them German) the impression that there existed a solid bloc opposing the Triple Alliance, there was no substance to this belief. No commitment, "moral" or otherwise, existed that would force Britain to intervene in any continental quarrel.[3]

It is also important to remember the Imperial aspects of Britain's position. Although several of the other European powers possessed substantial empires, none of these approached – in size, significance, and prestige – that of Britain. And, certainly, no other European power had an empire that posed such serious defense issues as did the British Empire.[4] For most of the nineteenth century, and, arguably, even until 1914, the defense of India was the major issue for those who dealt with British defense policy.[5]

The complexity of Imperial defense was compounded by how it was linked to European matters. This becomes clear if we compare Britain's possible involvement in a Europe war with its possible military involvement overseas. If Britain were to join a continental quarrel among the Great Powers, it would undoubtedly do so within the system provided by the balance of power. In this system, any attempt by a Great Power to alter the *status quo* on the Continent would automatically generate opposition, and Britain would have allies in any conflict.[6] Quite the opposite was true with respect to imperial wars. There, Britain would have to resolve matters utilizing only her own resources. And, as the South African conflict and the quarrels in 1898 with France at Fashoda and with Russia over China made evident, imperial involvements could both create opposition from the Great Powers and paralyze British foreign policy in Europe.[7]

[3] T. Wilson, "Britain's 'Moral Commitment' to France in August 1914," *History* 64 (1979): 380–90.

[4] The best studies are John Gooch, *The Plans of War: The General Staff and British Military Strategy c.1900–1916* (London, 1974); Aaron Friedman, *The Weary Titan: Britain and the Experience of Relative Decline, 1895–1905* (Princeton, 1988); and Rhodri Williams, *Defending the Empire: The Conservative Party and British Defence Policy 1899–1915* (New Haven and London, 1991).

[5] See Neilson, *Britain and the Last Tsar*, pp. 110–43, for the planning.

[6] T.G. Otte, "'Almost a Law of Nature'? Sir Edward Grey, the Foreign Office, and the Balance of Power in Europe, 1905–12," *Diplomacy and Statecraft* 14 (2003): 77–118. For a discussion of how this systemic issue played on another aspect of British strategic foreign policy, see Keith Neilson, "The Anglo-Japanese Alliance and British Strategic Foreign Policy, 1902–1914," in Phillips Payson O'Brien, ed., *The Anglo-Japanese Alliance, 1902–1922* (London, 2004), pp. 48–63.

[7] See two articles by Thomas Otte, "A Question of Leadership: Lord Salisbury, the Unionist Cabinet, and Foreign-Policy Making, 1895–1900," *Contemporary British History*

As Arthur Balfour, First Lord of the Treasury in Lord Salisbury's last government, commented during the Boer War, Britain was

> for all practical purposes at the present moment only a third-rate power; and we are a third-rate power with interests which are conflicting with and crossing those of the great powers of Europe. Put in this elementary form the weakness of the British Empire, as it at present exists, is brought home to one. We have enormous strength, both effective and latent, if we can concentrate . . . but the disposition of our Imperial interests . . . renders it almost impossible.[8]

Empire, clearly, was both a strength and a liability, with all that this implied for defense and war planning.[9]

The fact that Britain's Empire was a maritime one introduces another factor that makes British war planning different from that of the other Great Powers: the centrality and significance of the Royal Navy. For Britain, the Royal Navy played at least two important roles. First, it served as a barrier to invasion. While there were various efforts by the army to argue that it could prevent invasion – the "bricks and mortar" school of the nineteenth century and a number of attempts in the decade before 1914 to insist that preventing a "bolt from the blue" required an increase in British land forces – no one ever argued that the Royal Navy did not constitute the primary obstacle to any attempt to land a hostile force on British shores.[10]

Second, the Royal Navy had to establish command of the sea. For the British, this did not just mean, in the fashion that the term was used by such proponents of sea power as Alfred Thayer Mahan, that the Royal Navy would be able to defeat any enemy fleet. Instead, and following the views expressed by Julian Corbett, what the British required of the Royal Navy was to ensure that its seaborne commerce and transportation were able to move without effective opposition.[11] This was important for a number of reasons: Britain's economic and financial position was based

14 (2000): 1–26; and "Great Britain, Germany, and the Far Eastern Crisis of 1897–8," *English Historical Review* 90 (1995): 1157–79; and the useful contributions in Keith Wilson, ed., *The International Impact of the Boer War* (Chesham, 2001).

8 Hamilton (Secretary of State for India) to Curzon (Viceroy, India) 4 July 1901, quoting Balfour; Curzon Papers, MSS Eur F111/149, British Library.

9 See the discussion in Keith Jeffery, "The Eastern Arc of Empire: A Strategic View 1850–1950," *Journal of Strategic Studies* 5 (1982): 531–45.

10 John Gooch, "The Bolt from the Blue," in John Gooch, ed., *The Prospect of War: Studies in British Defence Policy 1847–1942* (London, 1981), pp. 1–34.

11 For his views, see Julian S. Corbett, *Some Principles of Maritime Strategy* (London, 1911). Corbett's work is examined in Donald M. Schurman, *Julian S. Corbett, 1854–1922: Historian of British Maritime Policy from Drake to Jellicoe* (London, 1981).

on exports; Britain imported 60 percent of its food supplies by 1914; and the defense of the British Empire required the rapid and safe transport of men and material by sea.[12]

The above considerations make it clear why, instead of talking about British "war plans," it is perhaps better to speak of British contingency plans, that is, to discuss what planning the British had done for the various kinds of war in which they might become involved. It suggests also that one should look not just at military thinking about a possible commitment of British forces to the Continent in the context of 1914, but also at both naval strategy and plans for the defense of the Empire.

We turn first to the issue of the defense of India. This matter was the dominant military issue in the second half of the nineteenth century and became particularly acute in the immediate aftermath of the Boer War (1899–1902). The difficulty of defending India was that the building of the Russian Trans-Siberian Railway, with a spur line from Orenberg to Tashkent, meant that Russian forces could threaten India more quickly than British reinforcements could be transported by sea.[13] Thus, the standing garrison in India would have to be large enough to respond to an initial attack and substantial reinforcements – 30,000 men within four months of the outbreak of hostilities and 70,000 further troops later – would have to be sent to the sub-continent.

By 1904, even these figures were deemed inadequate. Henry Horatio Lord Kitchener, the commander-in-chief of the Indian Army, insisted that the number of reinforcements needed was 135,614, a number that he continued to raise throughout the rest of the year. By early 1905, he suggested that 211,824 men would be needed between the eighth and fifteenth month after hostilities began. Such projections generated at least two major objections. The first was financial. Britain's Exchequer had

[12] For this, see David French, *British Economic and Strategic Planning 1905–1915* (London, 1982); and two articles by Avner Offer, "The Working Classes, British Naval Plans, and the Coming of the Great War," *Past and Present* 107 (1985); 204–26; and "Morality and Admiralty: 'Jacky' Fisher, Economic Warfare and the Laws of War," *Journal of Contemporary History* 23 (1988): 99–119.

[13] What follows is based on Neilson, *Britain and the Last Tsar*, pp. 126–32; and *idem*, "'Greatly Exaggerated': The Myth of the Decline of Great Britain before 1914," *International History Review* 13 (1991): 713–17. While Russia primarily used the threat to invade India as a means to exert diplomatic pressure on Britain, London took the threat very seriously if for no other reason than because any Russian activity on the North–West frontier might spark another rebellion in India on the lines of the Great Mutiny of 1857.

been shaken by the Boer War, and successive chancellors favored reducing, not expanding, defense spending. The other was the simple unavailability of men. Without conscription, such numbers as Kitchener argued were necessary could not be raised, and conscription was an unacceptable political option, despite having its advocates within Unionist circles.[14]

This meant that the defense of India could not be achieved by military means alone. Instead, the British had to base their contingency planning on a wider platform. This involved diplomacy. By way of a deterrent, the British insisted that the renewal of the Anglo-Japanese Treaty concluded in 1905 include a provision that Japan would provide troops to help defend India.[15] However, both the British and Japanese had second thoughts about this agreement. The British were concerned that using non-white troops to defend India might raise doubts in Indian minds about the innate superiority of Britons; the Japanese made it clear that, in any Anglo-Russian conflict, Tokyo would deploy troops in Korea rather than in the sub-continent.[16] More fruitful was the negotiation of an Anglo-Russian Convention, signed in 1907. The end of Anglo-Russian enmity meant that military planning for the defense of India lost much of its urgency. As a result, in the British deliberations before the Convention was concluded, it was decided that Indian defense would require 100,000 reinforcements, and the matter was left to rest there.[17] However, it is important to remember that the issue was only dormant, not resolved, and its continued slumber depended on the maintenance of good Anglo-Russian relations.[18]

14 R.J.Q. Adams and P. Poirer, *The Conscription Controversy in Great Britain 1900–18* (London, 1987), outlines the issues. There were Liberals who, when the war broke out, favored conscription; see Matthew Johnson, "The Liberal War Committee and the Liberal Advocacy of Conscription in Britain, 1914–1916," *Historical Journal* 51 (2008): 399–420.

15 See Ian Nish, *The Anglo-Japanese Alliance: The Diplomacy of Two Island Empires 1894–1907* (London, 1966), pp. 317–35.

16 Minutes, 84th meeting of the CID, 15 February 1906, Cab[inet Office] 2/2; "Anglo-Japanese Agreement of August 12, 1905. Proposals for Concerted Action. – Memorandum by the General Staff," CID-68B, November 1905, Cab 4/1. K.M. Wilson, "The Anglo-Japanese Alliance of August 1905 and the Defending of India: A Case of the Worst Scenario," *Journal of Imperial and Commonwealth History* 21 (1993): 334–56; for the context, see Neilson, "Anglo-Japanese Alliance."

17 "The Military Requirements of the Empire as Affected by the Defence of India," Morley (Secretary of State for India), 1 May 1907, Cab 16/2.

18 A glance at the debate over British military tactics and doctrine also underlines the fact that the army was preparing both for war in the Imperial and Continental contexts: see T.H.E. Travers, "Technology, Tactics, and Morale: Jean de Bloch, the Boer War, and British Military Theory, 1900–1914," *Journal of Modern History* 51 (1979): 264–86;

The latter was not a given, and, by the spring of 1914, the Anglo-Russian accord was on shaky grounds.[19] As northern Persia became increasingly Russified and as the tentacles of Russian railways snaked nearer to India, Delhi remained convinced that the Russian menace was recrudescing. Only the outbreak of war in 1914 prevented a complete deterioration of Anglo-Russian relations, and concerns about the eventual fate of India remained a staple of Imperial defense even after the beginning of hostilities.[20]

The defense of India was not the only issue for the British army. From the turn of the century, it also had an increasing interest in a continental commitment.[21] This resulted from two things: an elevated professionalization of the army as a consequence of the Boer War, the end result of which was the creation of a General Staff in 1905; and the gradual reorientation of British strategic foreign policy toward the possibility of intervention on the Continent.[22] This latter was pushed by a number of men in the War Office. The earliest of these was J.M. Grierson, who served as the initial (1905–06) Director of Military Operations (DMO) in the General Staff. Grierson was an admirer of the German army, and, as a member of the Intelligence Division of the War Office from 1889 to 1895, had made a number of trips to Berlin to study the German General Staff.[23] His next posting was to Berlin as military attaché, where he

Keith Neilson, "'That Dangerous and Difficult Enterprise': British Military Thinking and the Russo-Japanese War," *War and Society* 9 (1991): 17–37; and Howard Bailes, "Technology and Tactics in the British Army, 1866–1900," in Ronald Haycock and Keith Neilson, eds., *Men, Machines and War* (Waterloo, ON, 1988), pp. 21–49.

[19] Jennifer Siegel, *Endgame: Britain, Russia and the Final Struggle for Central Asia* (London and New York, 2002), pp. 143–96.

[20] Keith Neilson, "'For Diplomatic, Economic, Strategic and Telegraphic Reasons': British Imperial Defence, the Middle East and India, 1914–1918," in Greg Kennedy and Keith Neilson, eds., *Far Flung Lines: Studies in Imperial Defence in Honour of Donald Mackenzie Schurman* (London and Portland, OR, 1997), pp. 103–23.

[21] The phrase is Michael Howard's; see his *The Continental Commitment: The Dilemma of British Defence Policy in the Era of Two World Wars* (London, 1972). Particularly germane here are pp. 31–52.

[22] Halik Kochinski, "Planning for War in the Final Years of Pax Britannica, 1889–1903," in David French and Brian Holden Reid, eds., *The British General Staff. Reform and Innovation, 1890–1939* (London and Portland, OR, 2002), pp. 9–25, outlines the process by which the General Staff was created and introduces the literature. Hew Strachan's contribution in the same volume, "The British Army, its General Staff and the Continental Commitment 1904–14," pp. 75–94, informs what follows. The diplomatic changes are best followed in Zara Steiner and Keith Neilson, *Britain and the Origins of the First World War* (2nd ed, Basingstoke and New York, 2003).

[23] Thomas G. Fergusson, *British Military Intelligence, 1870–1914: The Development of a Modern Intelligence Organization* (London, 1984), p. 117.

remained from 1896 to 1900. During his time in Berlin, Grierson had become aware of a growing hostility toward Britain, and accordingly pushed for war plans to be made to address the possibility of a British expeditionary force being sent to the Continent.

When Grierson became DMO, the foreign policy context favored these leanings. Between 1900 and 1905, Anglo-German relations had, if not deteriorated, certainly not improved. British attempts to reach an agreement with Germany had been rebuffed by Berlin, whose asking price was too high for London to consider. Instead, the British had concluded the Anglo-Japanese Alliance in 1902 and the Anglo-French *entente* two years later. In 1905 the Germans attempted to sunder the latter by means of the first Moroccan crisis.[24] The new Liberal foreign secretary, Sir Edward Grey, realized the need to support France by military means if necessary, but it was clear that this aid was neither automatic nor permanent.[25] As he wrote to Sir Francis Bertie, the British ambassador to Paris, on 15 January 1906:

> Diplomatic support we are pledged to give [France] and are giving. A promise in advance committing this country to take part in a Continental war is another matter and a very serious one: it is very difficult for any British Govt to give an engagement of that kind. It changes the entente into an alliance and alliances, especially continental alliances are not in accordance with our traditions.[26]

Instead, Grey felt that it was "quite right that our Naval and Military Authorities should discuss the question in this way with the French."[27] And, by "in this way," Grey meant "unofficially and in a non-committal way."

It was in this context that Grierson and his successors as DMO, particularly Brigadier General Sir Henry Wilson (1910–14), continued to press their plans for a continental commitment. Wilson was a fervid advocate of such a course. An ardent Francophile and intriguer, the Ulsterman was

24 For the Moroccan crises, see F.V. Parsons, *The Origins of the Morocco Question 1880–1900* (London, 1976); E.N. Anderson, *The First Moroccan Crisis, 1904–1906* (Chicago, 1930); Keith Hamilton, *Bertie of Thame: Edwardian Ambassador* (London, 1990), pp. 69–124.

25 For Grey, see Keith Neilson, "'Control the Whirlwind': Sir Edward Grey as Foreign Secretary, 1906–16," in T.G. Otte, ed., *The Makers of British Foreign Policy from Pitt to Thatcher* (Basingstoke and New York, 2002), pp. 128–49.

26 Grey to Bertie, 15 January 1906, Grey Papers, FO 800/49, The National Archives (formerly the Public Record Office), Kew.

27 Grey to Tweedmouth (First Lord of the Admiralty), 16 January 1906, Grey Papers, FO 800/87.

perhaps the key individual in the General Staff who pushed for cooperation with the French and a continental commitment.[28] In 1906, 1907, and 1908 the British, in conjunction with the French General Staff, worked out detailed plans to send a British Expeditionary Force (BEF) to the Continent to help the French deal with the German menace.[29] And in October 1908 the Committee of Imperial Defence (CID), created in 1902 to serve as a forum in which British defense policy could be discussed by all the departments concerned with the formation of British strategic foreign policy – the Foreign Office, the Admiralty, the War Office and the Treasury – established a sub-committee to consider Britain's defense responsibilities beyond securing the Empire.[30] The sub-committee had as its chairman the prime minister, Sir Herbert Asquith. In the sub-committee, the General Staff put forward its plans for sending to the Continent a BEF consisting of four infantry divisions and one cavalry division (a number to be augmented by an additional two divisions when it was judged that the Territorial Force was ready to take over the responsibilities for Home Defence).[31]

This sparked heated opposition from the Royal Navy, which had its own vague plans for limited amphibious operations against Germany and argued that Britain's main contribution to any continental war would (and should) be maritime, not terrene. The strategic differences between the two services were not resolved.[32] Instead, a compromise was arrived at in which the form that a British contribution would take was left to be decided at the time of engagement. This allowed both the army and the navy to continue on their disparate paths without taking into considerations each other's plans.

[28] For Wilson, see Keith Jeffery, *Field Marshal Sir Henry Wilson. A Political Soldier* (Oxford, 2006), and C. Callwell, *Field Marshal Sir Henry Wilson: His Life and Diaries* (2 vols., London, 1927); for a perceptive account of his tendency towards intrigue, Brock Millman, "Henry Wilson's Mischief: Field Marshall [*sic*] Sir Henry Wilson's Rise to Power 1917–18," *Canadian Journal of History* 30 (1995): 468–86.

[29] Nicholas d'Ombrain, *War Machinery and High Policy: Defence Administration in Peacetime Britain 1902–1914* (Oxford, 1973), pp. 89–90.

[30] On the formation of the CID, see F.A. Johnson, *Defence by Committee: The British Committee of Imperial Defence, 1885–1959* (London, 1960); for the sub-committee, d'Ombrain, *War Machinery*, pp. 92–100.

[31] "Report of the Sub-Committee of the Committee of Imperial Defence on the Military Needs of the Empire," CID 109-B, 24 July 1909, Cab 4/3.

[32] John Gooch, "Adversarial Attitudes: Servicemen, Politicans and Strategic Policy in Edwardian England, 1899–1914," in Paul Smith, ed., *Government and the Armed Forces in Britain 1856–1990* (London and Rio Grande, 1996), pp. 53–74; especially, pp. 65–73. Also Jim Beach, "The British Army, the Royal Navy, and the 'Big Work' of Sir George Aston, 1909–1914," *Journal of Strategic Studies* 29 (2006), 145–68.

Such an approach was possible only when the international situation was not threatening. But that ended in 1911, when the Germans precipitated the second Moroccan crisis.[33] The result was that the British had to decide upon a possible course of action. This took place at a meeting of the CID on 23 August 1911.[34] For many historians, this meeting has taken on mythic proportions, marking the time at which the General Staff's strategy of a continental commitment became accepted British policy.[35] However, if the meeting is examined carefully, this is true only in a narrow sense.

The purpose of the gathering was to determine, in Asquith's words, how Britain might, if required, give "armed support to the French." To that end, first the army, and then the Royal Navy, outlined their proposals.[36] The exposition of the General Staff's plan, given by Henry Wilson and the Chief of the Imperial General Staff (CIGS), Field Marshal Sir William Nicholson, was nearly the same as that put forward in 1908; that is, a BEF consisting of six divisions (as the Territorial Force now was considered adequate for Home Defence) and a cavalry division should be dispatched to France in any war with Germany. However, it is the assumptions on which his recommendations were based that are of significance. The first was that "the policy of England is to prevent any one or more of the Continental Powers from attaining a position of superiority which would enable it or them to dominate the other Continental Powers." The second was that the Royal Navy's ability to effect this policy would come into play only in the long run.

Given these two assumptions, the General Staff considered two cases. First, if Britain were to remain neutral, then "Germany will fight France

[33] For that crisis, see J.-C. Allain, *Agadir 1911: Une crise impérialisme en Europe pour la conquête de Maroc* (Paris, 1976); Hamilton, *Bertie of Thame*, pp. 214–47; and Thomas Otte, "The Elusive Balance. British Foreign Policy and the French Entente before the First World War," in Alan Sharp and Glyn Stone, eds., *Anglo-French Relations in the Twentieth Century: Rivalry and Cooperation* (London and New York, 2000), pp. 23–26.

[34] 114th meeting of the CID, 23 August 1911, Cab 2/2. My account of the meeting is based on this document. The following quotation is drawn from it.

[35] Samuel R. Williamson, *The Politics of Grand Strategy: Britain and France Prepare for War, 1904–1914* (London and Atlantic Highlands, NJ, 1990), pp. 167–204; Gooch, *The Plans of War*, pp. 289–92. This argument has been widely accepted; however, Strachan's revisionist approach in "The British Army, its General Staff and the Continental Commitment," seems more convincing.

[36] "The Military Aspect of the Continental Problem. Memorandum by the General Staff.", CID 130-B, 15 August 1911, Cab 4/3; "The Military Aspect of the Continental Problem. Remarks by the Admiralty on Proposal (B) of the Memorandum by the General Staff (130-B)," CID 131-B, 21 August 1911, Cab 4/3. The following quotations are from these sources.

single-handed," as Russian help would not be forthcoming quickly enough to diminish German pressure in the West. The result was that "France would in all probability be defeated." The second case was if "England becomes the active ally of France." Again, the impact of Russia in the short term could be ignored. However, British intervention would both ensure that "the actual disparity in numbers becomes less" and "perhaps . . . the most important consideration of all . . . the morale of the French troops and nation would be greatly strengthened." The British intervention, to be effective, Wilson concluded, required that Britain's mobilization should occur as quickly as that of France and Germany and that the entire BEF be sent. "If we fail to do so, the value of our assistance would be seriously diminished." For the General Staff, the role of the Royal Navy in the immediate crisis was to ferry the BEF to France and to guard against any possible invasion.

None of these points cut much ice with the Admiralty. That body had its own plans for dealing with a German menace. The essence of this was to "blockade the whole of the German North Sea coast."[37] However, the Admiralty wished to combine this with a series of raids on the German coast that would enable British destroyers to "lie near to the shore" to effect a close blockade.[38] Such actions would require landing about a division of troops on German territory, along the Baltic Sea coast. The First Sea Lord, Sir Arthur Wilson, also argued that the General Staff's proposals would have serious drawbacks: a collapse of public morale in Britain "if the entire regular Army were dispatched abroad," an inability to assist the Royal Navy "in defence matters" in Britain, and a lack of "regular troops available for direct co-operation in the naval operations."

The Admiralty's plans, and its objections to the army's proposals, were subjected to intense criticism. The technical dangers of operating the fleet close to shore in any amphibious or close blockade actions were stressed.[39] The Admiralty also was reminded that in successive enquiries into the possibility of a German invasion, the Royal Navy had taken the

[37] Minutes of the 114th meeting, 23 August 1911, Cab 2/2. The following quotations are from this source.

[38] For a detailed look at the Admiralty's thoughts, see Paul Hayes, "Britain, Germany, and the Admiralty's Plans for Attacking German Territory, 1906–1915," in Lawrence Freedman, Paul Hayes and Robert O'Neill, eds., *War, Strategy, and International Politics: Essays in Honour of Sir Michael Howard* (Oxford, 1992), pp. 95–116.

[39] M. S. Partridge, "The Royal Navy and the End of the Close Blockade, 1885–1905: A Revolution in Naval Strategy?," *Mariner's Mirror* 75 (1989), pp. 119–36; Christopher Martin, "The 1907 Naval War Plans and the Second Hague Peace Conference: A Case of Propaganda," *Journal of Strategic Studies* 28 (2005), pp. 833–56.

position that Home Defence required only a limited number of British troops as the navy would prevent any larger German landing from taking place. Further, the Admiralty's plans for attacking the German coastal positions did not take into consideration the considerable advantages that land-based artillery had over naval gunfire, the likely need for lengthy sieges that the defense of Port Arthur in the Russo-Japanese War (1904–05) had shown, and the German ability to move reinforcements by rail to assaulted areas more quickly than any British reinforcements might be provided by sea. Even Maurice Hankey, the pro-Navy secretary to the CID, felt that the Admiralty's plan had been "cocked up in the dinner-hour."[40]

What emerged clearly from the meeting was that the Admiralty simply opposed any emphasis on sending a BEF to the Continent. In fact, the Admiralty went so far as to suggest that the Royal Navy would be unable, due to its own mobilization requirements, to provide the necessary transport and protection to allow the BEF to go to France immediately upon the outbreak of war.[41] This argument was disposed of when the War Office pointed out that the Admiralty had, in effect, agreed in 1908–09 that such assistance could be provided. While Hankey could optimistically contend that the "great thing is that we [the Royal Navy] are safe" and that "no decision was arrived at," in general it was clear that the Admiralty's alternatives were impracticable and even ill-considered.[42]

But, this did not mean that sending the BEF to the Continent had become British policy, and that the meeting of 23 August had "set the course for a military confrontation between Britain and Germany."[43] For example, in October 1911 Reginald Esher, an *éminence grise* in British defense planning, spoke with Asquith about defense matters:

> Then we talked about the General Staff scheme of landing and [*sic*] Army in France. The P.M. is opposed to this plan. He inclines to an occupation of Belgium

[40] Lord Hankey, *The Supreme Command 1914–1918* (2 vols., London, 1961), vol. I, p. 81. However, the Admiralty did not abandon its schemes entirely; in August 1914 a proposal to land Russian troops along Germany's Baltic coastline was bruited, and in 1915 Lord Fisher pressed for landings in Pomerania.

[41] For the arguments, see "The Oversea Transport of Reinforcements in Time of War," CID 116-B, 16 June 1910, Cab 4/3.

[42] Hankey to Admiral J. Fisher, 24 August 1911, as printed in Keith Wilson, "Hankey's Appendix: Some Admiralty Manoeuvres During and After the Agadir Crisis," *War in History* 1 (1994): 87.

[43] This point has been made by Strachan, "The Continental Commitment," pp. 77–78. The latter contention is from Niall Fergusson's deeply-flawed, *The Pity of War* (London, 1998), p. 65.

upon political and military grounds. He will not hear of the despatch of more than 4 Divisions. He has told Haldane [R.B. Haldane, the Secretary of State for War] so. He wants the Belgian plan worked out.[44]

Esher went on to contend to Asquith that the "mere fact" of the General Staff's having discussed detailed plans for the BEF's going to France had committed Britain "whether the Cabinet likes it or not." The prime minister did not accept such a view:

I asked the P.M. whether he thought that would be possible to have an English force concentrated in France within 7 days of the outbreak of war, in view of the fact that the Cabinet (the majority of them) have never heard of the plan. He thinks it impossible! How this would astound the General Staff.

The matter raised by this latter quotation both makes it clear how tentative was the "victory" of the General Staff in 1911 and underlines another significant point, the political opposition to any such war plans.

In the aftermath of the second Moroccan crisis, there was a wide-ranging attack on the foreign policy of Foreign Secretary Grey.[45] The Radical portion of the Liberal Party, irritated at having been excluded from the 23 August CID meeting and fearful that Britain was somehow committed to France, launched a "Cabinet cabal against the Entente with France and the Defence Committee."[46] This occurred at the Cabinet on 2 November, when Lord Morley, the Secretary of State for India, raised the "inexpediency of communication being held or allowed between the General Staff of the War Office & the General Staff of foreign States, such as France, in regard to possible military co-operation without the previous knowledge or direction of the Cabinet."[47]

This was dangerous ground, which threatened to split the government. Haldane outlined and explained the nature of the talks with France back as far as the initial conversations in 1906.[48] For his part, Asquith declared

[44] Esher journal entry, 4 October 1911, Esher Papers, ESHR 2/12, Churchill College Archives Centre, Cambridge.

[45] John A. Murray, "Foreign Policy Debated: Sir Edward Grey and his Critics, 1911–1912," in L.P. Wallace and W.C. Askew, eds., *Power, Public Opinion, and Diplomacy* (New York, 1959), pp. 140–71; Neilson, *Britain and the Last Tsar*, pp. 320–5.

[46] Esher journal entry, 24 November 1911, Esher Papers, ESHR 2/12.

[47] Asquith to the King, 2 November 1911, Asquith Papers I/6, Bodleian Library, Oxford. Germane, too, is Hankey's memorandum, likely prepared for this meeting, "The case against sending the Expeditionary Force to France," Hankey, November 1911, in Wilson, "Hankey's Appendix," pp. 87–98.

[48] This effectively quashes the notion that the Cabinet in 1914 was unaware of the existence of such talks; for a detailed account of this problem, see John W. Coogan and Peter F. Coogan, "The British Cabinet and the Anglo-French Staff Talks, 1904–1914: Who Knew What and When Did He Know It?," *Journal of British Studies* 24 (1985): 110–31.

that "all questions of policy have been & must be reserved for the decision of the Cabinet, & that it is quite outside the function of military or naval officers to prejudge such questions." As to what Britain's possible allies thought, the prime minister "believed (& Sir E. Grey concurred) that this was fully recognised by the French Government." Asquith's assurance did not completely satisfy the Cabinet, and the matter was returned to on 16 November. At that time, the foreign secretary made a statement designed to allay the Cabinet's fears:

> Sir E. Grey made it clear that at no step of our intercourse with France since Jan. 1906 had we either by diplomatic or military engagements compromised our freedom of decision & action in the event of war between France & Germany.[49]

However, this was not enough for many who felt that "communication of the kind referred to might give rise to [French] expectations, & that they should not, if they related to the possibility of concerted action, be entered into or carried on without the sanction of the Cabinet." The result was an agreement accepting this latter restriction and, more importantly, a declaration that "no communications should take place between the General Staff here & the Staffs of other countries which can, directly or indirectly, commit this country to military or naval intervention."

Grey had weathered the political storm and retained his control of British policy. However, it was clear that any "war plans" were tentative and subject to strict civilian control. While Grey declared that he did not want to "create consternation" by forbidding British military authorities from talking to the French, it was clear that such talks would have to be entered into with the Cabinet's permission and would be monitored closely.[50] A good indication of this were the Anglo-French naval talks of 1912. Designed as a balance to the Haldane Mission to limit British and German naval building, the Anglo-French conversations were scrupulously reported to the Cabinet. As Grey noted about the talks, "The Cabinet has now the same knowledge of what taken place that the Prime Minister & I have."[51] While the Anglo-French Naval Agreement carried with it an implication that the Royal Navy would protect the French Channel ports in time of war, this agreement was contingent upon a political decision to enter the war in the first place.

49 Asquith to the King, 16 November 1911, Asquith Papers I/6.

50 Grey to Asquith, 8 September 1911, Grey Papers, FO 800/100; Otte, "The Elusive Balance," pp. 24–5.

51 Grey's minute (5 May 1912) on Arthur Nicolson's confidential memorandum, 4 May 1912, Grey Papers, FO 800/94.

The poor showing of the Admiralty at the CID meeting of 23 August also had repercussions with respect to personnel. Both the First Lord of the Admiralty, Sir Reginald McKenna, and the First Sea Lord were replaced, the former by Winston Churchill and the latter by Admiral Sir Francis Bridgeman.[52] The latter did not hold the post for long; in December 1912 he was replaced by Admiral Prince Louis of Battenberg.[53] However, even in retirement, it was Sir John ("Jacky") Fisher, the dominant figure in the Edwardian navy and a former First Sea Lord, who had the ear of the new First Lord.[54] Churchill came to office determined, as always, to further his own career. In order to do this, he had to satisfy a number of different constituencies. On the one hand, he had to ensure the continuance of British naval supremacy; on the other hand, he had to do this within the parameters of the budget and without offending too many of his erstwhile Radical colleagues and supporters within the Liberal Party. Further, he had to deal with the shortcomings in the Admiralty that the CID meeting of 23 August had made manifest.

He did so by introducing a number of reforms, the most straightforward of which was the creation of a Naval War Staff, a body that could provide systematic planning to counter the arguments put forward by the War Office.[55] But, it was in the realm of naval strategy and tactics that Churchill made some radical decisions. To understand them, it is necessary to consider the emerging historiography surrounding the Royal Navy before 1914.[56] The traditional view of pre-1914 naval strategy was established by Arthur J. Marder, who, in keeping with the existing views of British foreign policy at the time, argued that Fisher initiated a "naval revolution" by introducing the *Dreadnought* and concentrating the Royal Navy in home waters to deal with the German menace.[57]

This view has been slowly undermined in the past decade. It has been shown that Fisher's motivations for concentrating the fleet were financial,

[52] On the latter, see Nicholas Lambert, "Admiral Sir Francis Bridgeman-Bridgeman (1911–1912)," in Malcolm H. Murfett, ed., *The First Sea Lords. From Fisher to Mountbatten* (London and Westport, CT, 1995), pp. 55–74.

[53] See John B. Hattendorf, "Admiral Prince Louis of Battenberg (1912–1914)," ibid., pp. 75–90.

[54] For this, see Nicholas Lambert, *Sir John Fisher's Naval Revolution* (Columbia, SC, 1999), pp. 244–46.

[55] For earlier planning, see Paul Haggie, "The Royal Navy and War Planning in the Fisher Era," *Journal of Contemporary History* 8 (1973): 113–31.

[56] What follows is informed by Nicholas A. Lambert, "Transformation and Technology in the Fisher Era: the Impact of the Communications Revolution," *Journal of Strategic Studies* 27 (2004): 272–97.

[57] Arthur J. Marder, *From the Dreadnought to Scapa Flow: The Royal Navy in the Fisher Era* (5 vols., London, 1961–70).

not strategic; that the introduction of the *Dreadnought* was part of this scheme and that Fisher preferred battle cruisers to dreadnoughts.[58] What Fisher intended was to use submarines and mines to create a flexible system of "flotilla defence" for narrow waters and to employ squadrons of battle cruisers, deployed at key stations worldwide, to ensure both that British commerce was safe from attack and to deny the seas to London's enemies.[59] Indeed, by 1914, the Admiralty had reached a decision to pursue economic warfare (as opposed to simple blockade), something that would have destroyed not only the economies of Austria-Hungary and Germany, but perhaps also those of the British Empire and thus much of the world.

Combined with this was the tactic of "plunges," taking advantage of Britain's sophisticated naval-industrial complex to introduce new naval technologies that would ensure that the Royal Navy's opponents were always one expensive research-and-development step behind their British counterpart. All of this was tied together by a sophisticated command and control system based on Britain's global cable system and emerging wireless capabilities that would both allow the battle cruisers to be directed wherever they were needed and keep the German fleet penned in narrow waters by submarines and torpedoes.[60] As Fisher put it, what was needed was "a fast Battle Cruiser of 30 knots like a spider in the middle of his web communicated with by smaller tramps (at £4 a ton!)

[58] Jon Sumida, *In Defense of Naval Supremacy: Finance, Technology and British Naval Policy, 1889–1914* (Boston, 1989); idem, "Sir John Fisher and the Dreadnought: The Sources of Naval Mythology," *Journal of Military History* 59 (1995), pp. 619–37; Charles Fairbanks, "The Origins of the Dreadnought Revolution: A Historiographical Essay," *International History Review* 13 (1991): 246–72; and Lambert, *Fisher's Naval Revolution*. I do not find convincing the arguments advanced by Matthew Seligmann in his article, "New Weapons for New Targets: Sir John Fisher, the Threat from Germany, and the Building of HMS *Dreadnought* and HMS *Invincible*, 1902–1907," *International History Review*, 30(2008): 303–31.

[59] Here, the work of Nicholas Lambert is essential. In addition to his *Fisher's Naval Revolution*, see his articles: "Admiral Sir John Fisher and the Concept of Flotilla Defence, 1904–1909," *Journal of Military History* 59 (1995): 639–60; "Economy or Empire: The Fleet Unit Concept and Quest for Collective Security in the Pacific, 1909–1914," in Kennedy and Neilson, eds., *Far Flung Lines*, pp. 171–89. Also important is Jon Sumida, "A Matter of Timing: British Battle Fleet Tactics in the Dreadnought Era, 1912–1916," *Journal of Military History* 67 (2003): 85–136; Partridge, "The Royal Navy and the End of the Close Blockade," pp. 119–36; and Peter F. Halvorsen, "The Royal Navy and Mine Warfare, 1868–1914," *Journal of Strategic Studies* 27 (2004): 685–707.

[60] In addition to Lambert, "Transformation and Technology," see also Paul Kennedy, "Imperial Cable Communications and Strategy, 1870–1914," *English Historical Review* 86 (1971): 728–52. The context is best followed in Daniel R. Headrick, *The Invisible Weapon. Telecommunications and International Politics 1851–1945* (Oxford, 1991), pp. 73–137.

fitted with wireless and covering the whole ocean in a net work of wireless communication who can tell the spider where the fly is and the bigger the fly the more placid the digestive smile as the spider eats them up. On every ocean a huge Cruiser in the centre of the net."[61]

When Churchill came into office, these ideas were never far from his mind, particularly as Fisher ensured that this was so.[62] There was, however, opposition to these radical concepts within the Admiralty. Arthur Wilson had never entirely shared Fisher's enthusiasms, and both Battenberg and Sir John Jellicoe, the latter the second-in-command of the Home Fleet, favored a mixed approach of capital ships and submarines over Fisher's exclusive reliance on new technologies. However, in 1914 Churchill was willing to take the bull by the horns, and divert monies from the construction of battleships to the building of the submarines and auxiliary craft necessary to put Fisher's vision into effect.[63] Thus, the outbreak of the World War I found the Royal Navy at the beginning of a transformation as to how it would achieve its strategic goal of establishing command of the sea.

The essential significance of this "revolution" in naval strategy for the purposes under discussion here is that it, far more than the plans discussed at the 23 August 1911 meeting of the CID, reflected the "war plans" of the Royal Navy. And, despite the internal debate over how and whether new technologies and approaches should be employed to ensure that the Royal Navy dominated the seas, there was no doubt that it intended to do so. The "defeat" of 23 August meant only that the Royal Navy's plans for employing, in Fisher's famous phrase, the army as a "projectile to be fired by the Royal Navy" would not be utilized at the outbreak of hostilities. If Britain were to give military aid to France, then it would employ the means put forward by the War Office, not by the Admiralty.

That, however, was a big "if." As debate in 1914 showed, it was by no means certain that Henry Wilson's well-honed plans would be put into practice. Nor, in fact, did it turn out that Wilson's plans survived

[61] Fisher to Rosebery, secret & private, "please burn," 10 October 1914, Rosebery Papers, MS 10124. National Library of Scotland.

[62] For an examination of Churchill's attitude toward the strategic threat posed by Germany, see John H. Maurer, "The 'Ever-Present Danger': Winston Churchill's Assessment of the German Naval Challenge before the First World War," in John H. Maurer, ed., *Churchill and Strategic Dilemmas before the World Wars* (London and Portland, OR, 2003), pp. 7–50.

[63] Nicholas Lambert, "British Naval Policy, 1913–1914: Financial Limitation and Strategic Revolution," *Journal of Modern History* 67 (1995): 595–626.

their first contact, not with the German enemy, but with Britain's political leaders. In the debate on whether to enter the war, the details of British war plans did not figure largely, although on 2 August the Cabinet agreed to protect the French north coast and Channel as a result of the Anglo-French Naval Agreement of 1912.[64] The fact that the Royal Navy had not, on Churchill's orders, stood down after its summer exercises, however, meant that there would be no conflict between the transport needs of the BEF and the Royal Navy's own requirements for mobilization.

The Cabinet's deliberations about entering the war centered on foreign policy and domestic politics. Those who made up the "Liberal imperialist" element in the Government – men like Grey, Asquith, and Haldane – insisted that Britain's international position required that London respond to the events on the Continent and threatened resignation. The more extreme Radicals opposed going to war. The bulk of the ministers wavered, unwilling to face the fact that events required a decision. Most of them were gradually won over to intervention by the fact that a decision for neutrality would bring down the Government. Here, the threat of a coalition government – as the Conservative leadership had pledged support for a declaration of war – also was a factor, since those who waffled had to consider the possibility that Asquith might form a government in which their posts were given to Conservatives. The result was that only two ministers resigned in protest once the decision to go to war was taken. "War plans" had not pushed the Cabinet's decision in any fashion.

The real challenge to the realization of the plans put forward in August 1911 came in the Cabinet once war had been declared on 4 August. The BEF's mobilization already was three days behind that of the French, and on 5 August a war council met at 10 Downing Street to consider Britain's options.[65] There was an immediate challenge to the plans that Henry Wilson and his staff had designed since 1911. Sir John French, the commander-in-chief of the BEF, argued that the three-day gap between the British and French mobilizations made all prewar plans obsolete, and suggested that the BEF be landed in Antwerp instead of at a French port, something that he had advocated as early as 1911. But this was quickly

64 My account of the July crisis is based on Steiner and Neilson, *Britain and the Origins*, pp. 229–57; Stephen J. Valone, "'There Must Be Some Misunderstanding': Sir Edward Grey's Diplomacy of August 1, 1914," *Journal of British Studies* 27 (1988): 405–24; and Keith Wilson, "Britain," in Keith Wilson, ed., *Decisions for War, 1914* (London, 1995), pp. 195–208. The following paragraph is also based on these sources.

65 Meeting of the War Council, 5 August 1914, Cab 42/1/4.

dismissed as impractical, as the Royal Navy could not promise to protect an expedition to Antwerp.[66]

French was not the only senior military man to suggest that prewar plans be abandoned. On 4 August Sir Douglas Haig, who was scheduled to command I Corps in France, wrote to the former Secretary of State for War, R.B. Haldane, and put forward an even more radical proposal. Haig suggested that the BEF should not go to France immediately: "This war will last many months, possibly years, so I venture to hope that our only bolt (and that not a very big one) may not suddenly be shot on a project of which the success seems to me quite doubtful – I mean the checking of the Germany advance into France." Haig preferred that Britain should wait "three months" before dispatching the BEF to the Continent, by which time it would "have quite a considerable army so that when we do take the field we can act decisively and dictate terms which will ensure a lasting peace."[67]

Haig's remarks did not affect the political decision-making process, but things did not go smoothly on that front. This reflected a change in personnel that altered the entire landscape of decision making. On 6 August, Lord Kitchener of Khartoum, Britain's pre-eminent Imperial soldier, was made Secretary of State for War.[68] This was the result of a series of unconnected events. Earlier in the year, the refusal of a number of British officers to act to suppress the Ulster Nationalists – the so-called "Curragh mutiny" – had led to the dismissal of the Secretary of State for War, Jack Seely.[69] The position had not been filled prior to the outbreak of war, with Asquith acting as his own War Secretary. In June 1914

[66] For French's thinking, see William J. Philpott, "The Strategic Ideas of Sir John French," *Journal of Strategic Studies* 12 (1989): 458–78. Neither of French's recent biographers do much to clarify French's ideas: Richard Holmes, *The Little Field-Marshal: Sir John French* (London, 1981); and George H. Cassar, *The Tragedy of Sir John French* (London, 1985). For a discussion of the Belgian option, see Keith Wilson, "The War Office, Churchill and the Belgian Option: August to December 1911," *Bulletin of the Institute of Historical Research* 50 (1977): 218–28.

[67] Haig to Haldane, personal, 4 August 1914, Haldane Papers, MS 5910, National Library of Scotland. I would like to thank Professor John Ferris for bringing this document to my attention. Haig's belief in a long war has not been recognized by his biographers. He and Kitchener were the first British commanders to realize the likelihood of a protracted struggle.

[68] The best studies of Kitchener are those of George H. Cassar, *Kitchener: Architect of Victory* (London, 1977); and *Kitchener's War: British Strategy from 1914 to 1916* (Washington, DC, 2004).

[69] For the Curragh incident, see Patricia Jalland, *The Liberals and Ireland: The Ulster Question and British Politics to 1914* (Brighton, 1980), pp. 207–47; I.F.W. Beckett, ed., *The Army and the Curragh Incident, 1914* (London, 1986).

Kitchener, British Agent and Consul-General in Egypt, returned to England on leave, scheduled to return to Cairo at the beginning of August. The fact that someone of Lord Kitchener's military experience, immense prestige and political rank – he had been made an earl in the June honor's list – was available made his selection as Secretary of State for War a seemingly perfect fit, despite his lack of political experience.

At the meeting of the War Council on 6 August, it became apparent that Kitchener did not share the beliefs that underpinned the General Staff's plans for war.[70] Unlike the General Staff, and the majority of military planners in all countries, Kitchener did not believe that the war would be short. Instead, he told the Cabinet that the war would last three years and require a multi-million-man army.[71] Kitchener also did not accept the idea that the entire BEF should go to the Continent. Worried about the political situation in Ireland, the Cabinet agreed that only four divisions should go to France initially, with the fifth to follow if and when circumstances warranted (something quite in line with Asquith's preferences in 1911) and the sixth was to be maintained as a strategic reserve. The disposition of this sixth division, in fact, became a *pomme de discord* between the British and French that lasted well into 1915.[72] Kitchener had at least two other concerns. First, he wished to ensure that the command of the BEF remained independent; his instructions to its commander, Sir John French, made this point clearly.[73] Second, the Secretary of State also worried about the possibility of invasion, and sent troops to the coast, disrupting the careful railway timetables for movement that the General Staff had worked out.

Nor did Kitchener share the prewar views about where the BEF should concentrate once it reached France. Determined that British forces should act as a reserve force (quite contrary to the French plans for the British), Kitchener did not want the BEF placed too far forward. And, as prewar

70 Meeting of the War Council, 6 August 1914, Cab 42/1/3.

71 There is a good discussion of these points in Roy A. Prete, "French Strategic Planning and the Deployment of the B.E.F. in France in 1914," *Canadian Journal of History* 24 (1989): 42–62. My account, unless otherwise noted, is based on Prete and Cassar, *Kitchener's War*, pp. 23–29.

72 Keith Neilson, "Kitchener: A Reputation Refurbished?," *Canadian Journal of History* 15 (1980): 207–22; William J. Philpott, "Kitchener and the 29th Division: A Study in Anglo-French Strategic Relations, 1914–1915," *Journal of Strategic Studies* 16 (1993): 375–407.

73 For this, see Richard Holmes, "Sir John French and Lord Kitchener," in Brian Bond, ed., *The First World War and British Military History* (Oxford, 1991), pp. 117–18. The relationship between Kitchener and French was an uneasy one.

British intelligence about German intentions was inadequate (and largely derived from French sources), Kitchener was unconvinced that the BEF should go to Maubeuge – as the French wished – and instead preferred to send it to Amiens.[74] This led to the immediate dispatch to London of emissaries from the French High Command. They urged Kitchener that the BEF should take up its agreed-upon position on the French left wing.[75] This special pleading resulted in Kitchener acceding to a concentration at Maubeuge, but this agreement did not take place until 12 August, further derailing the BEF's mobilization timetables. With the question of destination settled, the BEF finally moved, its new timetable for arriving on the French left wing fully six days behind its prewar estimates.

Mobilization, once agreed upon, proceeded smoothly.[76] To augment this, the British severed the German telegraph cables, ensuring both that German ships abroad were cut off from directives emanating from Berlin and that neutrals, particularly the United States, received their news about the war exclusively from British sources.[77] Southampton was the major port of troop embarkation; Le Havre the major port of disembarkation. The Fifth, Seventh, and Eighth Battle Squadrons of Vice Admiral Sir Cecil

[74] On the failure of pre-1914 British intelligence to discern German intentions, see Fergusson, *British Military Intelligence*, pp. 197–232; Nicholas P. Hiley, "The Failure of British Espionage Against Germany, 1907–1914," *Historical Journal* 26 (1983): 867–89; and David French, "Spy Fever in Britain, 1900–1915," *Historical Journal* 21 (1978): 355–70. For the state of British knowledge of German capabilities and intentions (as held by the military and naval staffs), see Matthew S. Seligmann, *Spies in Uniform. British Military and Naval Intelligence on the Eve of the First World War* (Oxford, 2006). The influence of such information is difficult to assess. For example, while one of the prewar British military attachés strongly believed that Germany was planning war against Britain, his reportage was largely discounted; see Matthew S. Seligmann, "A View from Berlin: Colonel Frederick Trench and the Development of British Perceptions of German Aggressive Intent, 1906–1910," *Journal of Strategic Studies*, 23 (2000): 114–47. The state of British intelligence on the outbreak of war and its development can be found in Jim Beach, "'Intelligent Civilians in Uniform': The British Expeditionary Force's Intelligence Corps Officers, 1914–1918," *War & Society* 27 (2008): 1–22.

[75] This was the first in a long line of French attempts to persuade (or coerce) the British to follow in Paris's strategic footsteps; for an introduction to the difficulties, see William Philpott, "Britain and France Go to War: Anglo-French Relations on the Western Front 1914–1918," *War in History* 2 (1995): 43–64.

[76] The following from *History of the Great War, Naval Operations* ed. Sir Julian Corbett (London, 1920–31), vol. 1, pp. 72–82.

[77] Jonathan Reed Winkler, *Nexus: Strategic Communications and American Security in World War I* (Cambridge, MA, 2008), pp. 5–33. Also important is the same author's article, "Information Warfare in World War I," forthcoming in the *Journal of Military History*. I would like to thank Dr.Winkler for allowing me to read and refer to an advance copy of this piece.

Burney's Channel Fleet closed both ends of the English Channel against enemy raids, while French destroyers and submarines of the Boulogne Squadron guarded the Straits of Dover. The Grand Fleet was kept at sea to intercept any attempt by the German High Seas Fleet to attack British transports. There were neither formal convoys nor escorts. The 130,000 soldiers of the BEF proceeded in thirty-four ships to France, either singly or in pairs, by day and by night. Not a single troopship was lost due to enemy action.

Moreover, British naval plans initially went smoothly in 1914. The arguments between those who favored an all-out naval clash in the North Sea and those who advocated blockade were resolved by the German unwillingness to seek battle. As a consequence, the Royal Navy initiated its "information war" against German maritime strength, both naval and commercial.[78] Utilizing the Naval Intelligence Department's Trade Division's extensive data, collected worldwide and communicated to a War Room in London, the Royal Navy was able to maintain a close eye on German ship movements.[79] Despite some initial setbacks, by the end of 1914 the Admiralty was able to achieve an effective blockade and to clear the world trade routes of German commerce raiders. Neutrals were co-opted or bullied as appropriate.[80] *Der Tag*, the clash of the British and German fleets, would have to wait until 1916 and Jutland. The Royal Navy's plans for assaults on the German coast were never abandoned; however, amphibious operations instead became focused on other venues, the Dardanelles and the Persian campaign.

It is important to note that the Admiralty's plans for economic warfare – as opposed to blockade – had to be shelved. During August, the initial phases of economic warfare were begun, but these proved to be counterproductive. There were extensive complaints, not only from the neutrals, particularly the United States, but also from British merchants

78 For the concept of an "information war," see Joseph S. Nye, Jr., and William A. Owens, "America's Information Edge," *Foreign Affairs* 75 (1996): 20–36. This should be seen in the context of "knowledge-based organizations"; see N. Stehrs, *Knowledge Societies* (London, 1994), pp. 91–119.

79 For this, see Nicholas Lambert, "Strategic Command and Control for Maneuver Warfare: the Naval Intelligence Department and the Creation of the 'War Room' Plot, 1905–1915," *Journal of Military History* 69 (2005): 361–410. I would like to thank Dr. Lambert for allowing me to read an earlier version of this article. Also important is Avner Offer, *The First World War: An Agrarian Interpretation* (Oxford, 1989), pp. 226–32.

80 For an example, see Marc Frey, "Trade, Ships and the Neutrality of the Netherlands in the First World War," *The International History Review* 19 (1997), 541–62.

and shippers whose entire livelihood was being destroyed. Such complaints proved effective. The Admiralty's plans for economic warfare were deemed likely to produce only a Pyrrhic victory, and, instead, an economic blockade, controlled and coordinated by the Foreign Office, was put into place.[81] While the Admiralty continued throughout the war to kick at the controls placed on it by political concerns and to contend that the Foreign Office was "soft" on those who violated the blockade, it was evident that the requirements of domestic politics, the need to assuage neutral opinion and the necessities of alliance politics trumped the narrow thinking about economic warfare that was prevalent at the Admiralty.[82]

What does the above suggest about British "war plans"? First, it points out the extent to which they remained subservient to civilian decisions. Despite the efforts of such men as Sir Henry Wilson, the General Staff's plans never became an automatic response to a crisis. This resulted from a number of things. Perhaps most significant was Britain's detached diplomatic stance. Standing outside any formal alliances, the British had no formal obligations to either France or Russia. As a result, all of London's war plans were contingent upon political decisions about whether Britain should go to war. With a Liberal Party possessing a large pacifist (or, at best, neutralist) rump, British war plans would be carried out as a result only of a political decision and would not become a factor in the political decision itself. Second, Britain's position as an Imperial and maritime power added a complication to war plans generally, and another possible dimension to Britain's participation in any continental war. Even within the army, there was no consensus that British war plans and doctrine should be directed exclusively towards war on the Continent. And, reflecting the strength of the Royal Navy and Britain's own past, there were those who advocated that Britain largely concentrate on the

[81] For some instances of the problems involved, see Phillip Dehne, "From 'Business as Usual' to a More Global War: The British Decision to attack Germans in South America during the First World War," *Journal of British Studies* 44 (2005): 516–535; Maartje M. Abbenhuis, "In Fear of War: The First World War and the State of Siege in the Neutral Netherlands, 1914–1918," *War in History* 13 (2006): 16–41.

[82] For this and an introduction to the vast literature on blockade, see three articles by Greg Kennedy: "Intelligence and the Blockade, 1914–17: A Study in Administration, Friction and Command," *Intelligence and National Security* 22 (2007): 699–721; "The North Atlantic Triangle and the Blockade, 1914–1915," *Journal of Transatlantic Studies* 6 (2008): 22–33; "Strategy and Power: The Royal Navy, the Foreign Office and the Blockade, 1914–1917," *Defence Studies* 8 (2008): 190–206. These articles highlight the need for a new study of the blockade.

maritime aspects of a war, utilizing its limited military forces in amphibious actions and employing the Royal Navy in its traditional role to maintain command of the sea. Not only did the Government have to decide in a time of crisis whether Britain would enter into a continental war, it also had to decide upon the nature and extent of Britain's contribution. British war plans were an instrument of policy, never something that drove policy itself.

All of this was evident in August 1914. The British declaration of war was not driven by the dictates of war planning. Nor, when the commitment to arms was made, did the war plans automatically reduce the significance of civilian decision making. The impact of individuals remained supreme, as Kitchener's tinkering with the General Staff's plans made clear. The British had no "war by timetable"; instead, they had what might be termed "war by Cabinet agenda."

7

Italy

John Gooch

The kingdom of Italy, legally created in 1861 and territorially united nine years later with the capture of Rome, carried into the early twentieth century legacies of unification which impacted upon the process of military planning in both general and specific ways. The wars of the Risorgimento left in their wake an enmity toward Austria that was matched by a fear of and distaste for France. Both periodically resurfaced according to the turns of international politics. The political architecture of Liberal Italy created a governmental system with no functioning cabinet structures to require, encourage, or facilitate much in the way of coordination between the two services or between them and the bearers of the civil portfolios. The military were regarded by civilian politicians as at most a functional elite – though to Giovanni Giolitti, whose premierships dominated the first years of the twentieth century, they were barely functional and certainly not an elite. Of questionable utility as a weapon in foreign policy, the army was commonly perceived by both the Left and the Right as primarily an instrument for domestic coercion in the face of growing social agitation. In the circumstances, it is scarcely surprising that military planning before 1914 was in most ways a highly circumscribed activity into which issues such as the balance between military service and civilian economic activity seem never to have entered. Italy's soldiers looked to win battles – or at least not to lose them – on the assumption that this would win whatever war the politicians decided to start.

Perceived Threats and National Weaknesses

The planning culture that prevailed in the Italian General Staff was governed by a number of suppositions. Perhaps the most important was the perception of Italy's vulnerability, which was both internal and external. Internally the strength of Italian regionalism generated a deep sense of insecurity that was strengthened as social agitation increased during the last decade of the nineteenth century, boiling up into an open insurrection in Milan in May 1898.[1] Thanks in part to the growing strength of socialism in Italy, bouts of considerable internal disorder continued up to the very eve of the war. Externally, Italy's membership of the Triple Alliance from 1882 formally directed her strategic attention towards France, a focus of political hostility that partly derived from her role in the Risorgimento, partly from her role as a rival power in the Mediterranean and partly from economic competition. Although Austria-Hungary was formally an ally, strong undercurrents of irredentism which King Vittorio Emanuele II and his successor, Umberto I, publicly but somewhat fruitlessly disavowed, made the relationship an uneasy one and the potential threat of interference on behalf of the Pope by Catholic Vienna periodically alarmed an anti-clerical Rome.[2]

For political and strategic reasons, as well as economic ones, preventive war never figured in Italy's military lexicon and neither did it do so in the directing circles of politics and diplomacy – other than briefly under the premiership of the excitable Francesco Crispi in 1887–8. Militarily, the problem of defending Italy's unruly and, given the dearth of railways, distant islands, her extensive coastline and her porous Alpine mountains presented what seemed to be nightmarish difficulties. In a war with France, the planners had to face the possibility of separate campaigns on all three fronts; in a war with Austria, campaigns on two (the northeastern frontier and the Adriatic). In both cases, the evaporation of Swiss neutrality would increase the complexity of the problem. The most difficult task for the Italian General Staff was therefore that of prioritizing between multiple and complex scenarios. It was one they never solved. To the task of grappling with it they brought a well-established

[1] John A. Thayer, *Italy and the Great War: Politics & Culture, 1870–1915* (Madison and Milwaukee, 1964), pp. 30–1, 33; Umberto Levra, *Il colpo di stato borghesia: La crisi politica di fine secolo in Italia 1896/1900* (Milan, 1975), pp. 102–5.

[2] Federico Chabod, *Storia della politica estera italiana dal 1870 al 1896* (Bari, 1962), pp. 10, 37–8, 48,275, 448–9, 474, 668–76.

and deeply rooted, defensively minded outlook which military plans both accommodated and reflected.[3]

As a member of the Triple Alliance from 1882, Italy formally lined up with Germany and Austria-Hungary and geography dictated that Italian forces would most likely fight France in any war. Three strategic options would present themselves over the years: an attack across the common frontier of the Alpes Maritimes, which posed considerable problems; trans-shipping Italian troops to the Rhine via Austria or possibly neutral Switzerland to fight alongside the German army; or (briefly) launching an amphibious operation to land troops on the southern coast of France. The twists and turns of Italy's diplomatic posture produced a fourth scenario. At the start of the twentieth century, animosity towards France declined just as suspicion of Italy's other ally, Austria-Hungary, grew. War planning then had to accommodate a defensive campaign on the plains and in the foothills north and east of the Po valley.

A further influence on war planning was the prevailing assumption about the shape that war was going to take and the likely characteristics of future combat. The infantry regulations of 1883, 1885, and 1903 all envisaged frontal and flank attacks; those of 1910 stressed controlled and coordinated offensives and suggested that the target should be the enemy's lines of communication; and the 1913 combat regulations stressed the importance of firepower (provided primarily by artillery and rifles, not machine guns), movement and speed as the central ingredients both in open warfare and in attacking prepared positions.[4] By 1911, the chief of the General Staff believed that "every war will start with frequent operations of attack and defense against reinforced positions surrounding a central organized nucleus."[5]

Bureaucratic practice also contributed heavily to the shaping of planning culture. The dictates of managing a conscript army, when combined with national geography and the deficiencies of national communications, made the mobilization, assembly and deployment of troops a complex and difficult task; thus the construction of movement schedules and railway timetables took up a large part of the planning effort. War

[3] John Gooch, "Italy before 1915: The Quandary of the Vulnerable," in Ernest R. May, ed., *Knowing One's Enemies: Intelligence Assessment before the Two World Wars* (Princeton, 1984), pp. 205–33.

[4] Filippo Stefani, *La storia della dottrina e degli ordinamenti dell'esercito italiano* (Rome, 1984), vol. I, pp. 384–5, 411, 435, 446, 448, 458–76.

[5] Archivio dell'Ufficio Storico dello Stato Maggiore dell'Esercito (hereafter AUSSME). Relazione del viaggio dei generali 1911, p. 119. Rep. G-28 racc. 44.

plans – *progetti di Guerra* – combined an outline war scenario with the identification of a geographical theater of operations and proposals for the optimum deployment of the forces available within that theater. Projections of the shape of combat beyond the opening of hostilities either did not figure in the plans as such or else amounted to little more than the identification of lines of possible advance towards a limited number of physical objectives. This was a "functional" conception of planning, in which the army developed options for the employment of military force while the broader aspects of going to war which involved political considerations and decisions on the provision of the material to be used in carrying them out were left entirely to the political authorities. It persisted beyond the world war and into the Fascist era.[6]

Economics ensured that Italy was militarily as well as diplomatically "the least of the Great Powers." Its defense expenditure was less than that of all the other great powers except where navies were concerned; there, only Austria-Hungary spent less than Italy did. The need to balance the budget and the consequences of a tariff war with France, which lasted until 1898, meant that money for the army was tight during the last decade of the nineteenth and the first decade of the twentieth centuries. Even so, however, at approximately 250 million lire a year, the military budget took 40 percent of total state expenditure in 1901. Between 1907 and 1912, annual military expenditure averaged 511 million lire, now a quarter of state spending, and in 1912–13 it rose to 837 million, not including the costs of the Libyan war. Libya proved financially disastrous, taking 2 billion lire between 1911 and 1913, which amounted to just over 70 percent of the total annual state budget and leaving Italy with empty magazines and unreplenished stocks on the eve of the world war.[7]

In comparative terms, the numbers were very much against Italy: at £13.5 million a year, its average annual expenditure on the army in the decade 1904–14 was two-thirds what Austria-Hungary spent and half that of Great Britain. France spent two-and-a-half times as much as Italy, Germany three and a half times, and Russia four times as much.[8] For this reason, Italy was unable to do much more than redeploy her extant manpower when it faced direct military competition with Austria-Hungary

6 Fortunato Minniti, "Piano e ordinamento nella preparazione italiana alla Guerra degli anni trenta," *Dimensioni e problemi della ricerca storica* 1, 1990, pp. 132–3.

7 Giorgio Rochat and Giulio Massobrio, *Breve storia dell'esercito italiano dal 1861 al 1943* (Turin, 1978), pp. 153, 162–3.

8 David Stevenson, *Armaments and the Coming of War: Europe 1904–1914* (Oxford 1996), p. 8 table 8.

between 1906–08, and lost ground when the European powers first reequipped their armies with quick-firing artillery and then entered an arms race spiral in 1912–13.[9] The fact that Austria-Hungary started improving its frontier fortifications from 1912 sent shivers of anxiety down the corporate military spine in Rome.

Planners and Planning in the *Regio Esercito*

The contrast between the Italian failure at Custozza and the Prussian success at Sadowa in 1866 was evidence that the German model could not be ignored, and in imitation of the Berlin *Kriegsakademie* the *Scuola Superiore* opened its doors to staff officers in Turin on 11 March 1867, transmuting in September 1875 into the *Scuola di Guerra*. Although this institution produced graduates trained to act as executive staff officers with the troops, the higher functions of a staff were not being exercised by anyone. The Franco-Prussian War (1870–71) resulted in the creation of a pale imitation of the German General Staff in 1873, this in the form of a consultative body of general officers whose function was to advise on great military questions. Its impact was seemingly nugatory for two years later Emory Upton, in his survey of the structures of European armies, reported that the Italian army had no planning agency at all.[10] The modern staff system – and with it the beginnings of modern war planning – was ushered in with the appointment of General Enrico Cosenz as first army chief of staff on 3 August 1882.

The first two chiefs of staff, Enrico Cosenz (1882–1893) and Domenico Primerano (1893–1896), were both southerners who had been educated at the military college of the Nunziatella in Naples and initially destined for the Bourbon army. Both men had impeccably nationalist credentials and had fought (in Cosenz's case with great distinction) in the campaigns of the Risorgimento. Cosenz was acknowledged to be one of the most studious as well as one of the most modest men in the army. However, the fact that the new post was not seized by the Piedmontese clique at a time

[9] David Herrmann, *The Arming of Europe and the Making of the First World War* (Princeton, 1996), pp. 59, 108, 173–98; Stevenson, *Armaments and the Coming of War*, pp. 104, 163–4.

[10] Emory Upton, *The Armies of Asia and Europe: embracing official reports on the armies of Japan, China, India, Persia, Italy, Russia, Austria, Germany, France, and England* (New York, 1968[1878]), pp. 99–100. At that time the war department staff consisted of statistical, historical, railway, information, and topographical sections and a military education bureau.

when tensions between the two factions were only beginning to decline suggests that it was not yet seen as being of any great importance. Study of the wars of 1866 and 1870–71 convinced Cosenz that the secret of German success lay in the dominating direction of the military machine imparted by Helmuth von Moltke (the Elder), the quality of the Prussian army and the studied preparation that characterized its approach to war.[11]

The organizational structure of the higher reaches of the Italian army, however, dictated in part by the constitutional position occupied throughout the period by the king, meant that the chief of the Italian General Staff was not able to develop the degree of autonomy achieved by his German opposite number. The royal decree that defined the new post in July 1882, and that remained for the next twenty-four years, gave the chief of staff "the higher direction of studies for the preparation of war" in peacetime under the authority (*dipendenza*) of the minister for war, to whom he was to make proposals regarding the war formation of the army and with whose agreement he established "the general norms for mobilization and deployment plans according to the various war hypotheses."[12] In carrying out the duties of his post, Cosenz appears to have been both an austere and an isolated figure, said by the general staff at headquarters to be less accessible than the Pope.

Domenico Primerano, who succeeded Cosenz on 3 November 1893, had a short and troubled tenure and was left entirely outside the planning loop by the premier, Francesco Crispi, and the war minister, Stanislas Mocenni, during the final stages of the disastrous war in Ethiopia. The fact of his isolation could not be disguised in the subsequent parliamentary inquest and was pointed out in the Chamber of Deputies shortly after his resignation on 14 May 1896, when Foreign Minister Giulio Prinetti remarked that the government's green book of documents published after the debacle at Adua contained no allusion to him at all.[13] That debacle, on 1 March 1896, triggered an intense but short-lived public debate in which the proponents of something akin to a German system – a position which had earned them the uncomplimentary soubriquet "counterfeit Prussians" over the years – argued that the chief of the General Staff was

[11] Massimo Mazzetti, "Enrico Cosenz, scrittore militare'" *Il pensiero di studiosi di cose militari meridionali: Atti del Congresso* (Rome, 1978), pp. 100, 103–5.

[12] "Il Capo di Stato Maggiore," *L'esercito italiano* 15 September 1882. See also Filippo Stefani, *La storia della dottrina e degli ordinamenti dell'esercito italiano* (Rome, 1984) vol. I, pp. 311–2.

[13] "La dimissione del Capo di Stato Maggiore," *L'esercito italiano* 17 May 1896.

the natural head of the army and should be completely independent of the war minister. However, the latter's wide political responsibilities to Parliament and his position as a minister of the crown to whom the king's constitutional authority as commander of the army under the Statuto of June 1848 had been delegated, meant that no Italian chief of the General Staff could ever enjoy the independence of a Moltke.[14]

The appointment of Lieutenant-General Tancredi Saletta, a Piedmontese artilleryman from Turin and a soldier with combat experience in Africa who had previously served as Primerano's deputy, as chief of staff (1896–1908) signaled a recognition of both the importance and the potential sensitivity of the post. Saletta, a highly competent professional soldier with no shred of political ambition and a predilection for attending to the most minute of details, had his hands full: as well as supervising the preparation of new mobilization and mustering plans, publishing a new generation of service regulations, reestablishing grand maneuvers in 1903 and organizing them himself, he introduced staff rides akin to those conducted in Germany for some time past.

It was almost certainly largely as a result of the way in which Saletta carried out his duties in post that the government felt able to make important modifications in the status of the chief of the General Staff by royal decree in March 1906 and March 1908. Gaining a greater degree of power vis-à-vis the war minister, he was given authority to prepare *progetti d'operazione* to be undertaken after the assembly of troops at deployment, the point beyond which his predecessors had been debarred from trespassing as this marked the moment when tradition dictated that the field commander took direction of the war. The government was now required to keep the chief of the General Staff up-to-date with the politico-military situation. The chief of staff was also given charge of troop training and the drawing up of regulations for the tactical employment of large units, as well as the right to present projects "necessary or convenient for war" which had to do with military laws, regulations, or budgets.[15]

As the first Piedmontese officer to occupy the position, and as an official whose practical capacity and unconditional political loyalty had been rewarded when the king appointed him to the Senate in 1900, Saletta

[14] "Ministro della guerra e capo dello stato maggiore," *L'Italia militare e marina* 13–14 October 1896; "Uomo politico o tecnico?," *L'Italia militare e marina* 22–3 January 1900.

[15] "Il Capo di Stato maggiore dell'esercito," *L'Italia marina e militare* 3–4 April 1906.

had demonstrated that the occupant of his post was someone the highest circles in the army could trust with a degree of independence they had been unwilling to share with his predecessors. In practice, however, the gain was a limited one, as Alberto Pollio (1908–1914), the fourth chief of the General Staff, found out. In reality he was no more independent of the War Ministry – or of the crown – than his predecessor and no better incorporated into the counsels of government, which in Liberal Italy remained small and obdurately self-contained.

In 1910 the war minister, General Paolo Spingardi, proved markedly unwilling to alter or expand the authority of the chief of General Staff, holding to the traditional view that occupants of his office had maintained for several decades, namely that the *Capo supremo* delegated his authority to the war minister. A final royal decree two years before the world war broke out clarified the position in wartime, but only to a degree – when combat began the chief of staff would interpret the thoughts of the *Comandante supremo* and translate them into orders.[16]

The scope for Italian military planning, and its nature, was thus shaped to a large extent by the delicate position of the chief of General Staff as one member of a group which also included the king, the prime minister and the minister for war, a situation that greatly constrained the planning process because the responsibility for military policy was never clearly divided between them before the outbreak of war in August 1914.[17] In peacetime his subordination (*dipendenza*) to the War Ministry constrained his freedom, while the uncertainty as to who would command in war time hindered planning beyond the stage of deployment. All five of the prewar chiefs of staff thought that in war they should be the *Comandante superiore*, but nothing was firmly decided until Vittorio Emanuele assumed titular command of the armies on 26 May 1915 and nominated General Luigi Cadorna, who had succeeded to the post of chief of the Italian General Staff almost a year earlier, in July 1914, to translate his orders into actions.

How well the General Staff was manned depended heavily on the education provided by the *Scuola di guerra*, three-quarters of whose graduates transferred into the General Staff after passing their final

[16] "Il Ministro borghese e il Ministero della Guerra," *L'esercito italiano* 31 December 1907; "La difesa dello stato e l'opera del ministero della guerra borghese," *L'Italia militare e marina* 6–7 February 1908; "Il Capo di Stato Magg. dell'Esercito," *L'esercito italiano* 11 May 1910.

[17] Giorgio Rochat, "L'esercito italiano nell'estate 1914," *Nuova Rivista Storica* XXXXV, 1961, p. 31.

examinations, a proportion fixed in advance. Cosenz wanted more time spent on military history as happened in the Berlin *Kriegsakademie*, but lost the argument to the commandant, General Carlo Corsi, who suppressed the third-year course in military literature and replaced it with an obligatory social science course.[18] Cosenz would have preferred a two-year course to the three-year model in operation, but recognized that until the teaching in the military preparatory schools improved it would be difficult to produce officers whose level of education was high enough to cope with a more intensive course.

Another problem that beset the General Staff during this period was the shortage of candidates for the *Scuola di guerra*. In 1885, only sixty-eight candidates sat the initial written examination of whom thirty subsequently passed the orals and gained entry – a figure which remained roughly constant throughout the decade. Concerned at the inadequacy of the pool from which his candidates came, the chief of the General Staff suggested promotion by selection to the rank of major for all those passing out of the staff college as a way of encouraging more interest.[19] Under the influence of many of its senior commanders, the officer corps as a whole, from whom the candidates for the General Staff were drawn, remained resolutely anti-intellectual throughout most of the period.[20]

The General Staff was relatively small in number – 137 officers from 1889 until after 1908 – but life as a member does not seem to have been particularly arduous. Eugenio De Rossi, who served in the historical section between 1900 and 1903, recorded that a large part of the literary work in which he was engaged consisted in writing declarations and edicts in foreign languages directed at the inhabitants of hypothetically invaded lands. The three officers who manned the transport department spent their days smoking, laughing, and talking while one, who was training to be a singer, periodically burst into song.[21] However, a more authoritative

[18] Archivio Centrale di Stato (hereafter ACS). N.376/167, Cosenz to Ricotti, 19 January 1885; n. 376/1750, Cosenz to Ricotti, 22 August 1885. Ministero della Guerra: Segretario Generale: Scuole Militari, b. 2.

[19] ACS. N.34/3329, Cosenz to War Ministry; [unnumbered] Cosenz to War Ministry, 14 July 1887. Ministero della Guerra: Segretario Generale: Scuole Militari, b. 24. In 1888, 55 candidates took the written examination, 38 took the orals, and 28 passed into the staff college.

[20] Emilio De Bono, *Nell'esercito nostro prima della guerra* (Milan, 1931) p.197; Felice De Chaurand de St. Eustache, *Come l'esercito nostro entro' in guerra* (Milan, 1929), p. 133.

[21] Eugenio De Rossi, *La vita di un ufficiale italiano sino alla guerra* (Milan, 1927), pp. 80–1.

commentator noted that reinvigoration of the *Scuola di guerra* had led to "incalculable progress" in the General Staff by 1903.[22]

By the end of the nineteenth century, the General Staff was unpopular with the army at large. Marshal Emilio de Bono, writing during the Fascist era, described a corps of the *bel, biond, nobil e cirula* ("handsome, fair-haired, aristocratic and blue-eyed") who lived in the clouds, out of touch with ordinary mortals, and who skillfully avoided rotation away from headquarters in Rome to the rigors to be found in the territorial commands across the country.[23] The cause was to be found in the privileges which were becoming its particular perquisites. Accelerated promotion by selection, established when a decree in 1882 gave officers serving on the General Staff faster promotion, caused uproar when statistics showed that General Staff captains were being promoted with between four and seven years less seniority than their counterparts in all the other arms.[24] Granting all captains passing the *Scuola di guerra* the same privilege in 1893 did not improve the staff's popularity: artillery and engineer officers, who predominated there, were sent on graduation to cavalry or infantry battalions where they blocked promotion for those lower down, an increasing problem as military budgets were cut during the 1890s and drastic economies introduced.

The problem of the relationship between the staff and the rest of the army was so deeply entrenched that in 1904 a parliamentary commission claimed that the General Staff dominated the army from the rank of major upwards. Charging it with being a closed caste with a monopoly on fast promotions that excluded officers who did not come from its own ranks, the commission recommended abolishing it altogether in order to raise morale. The charges were unfair in that competition for the *Scuola di guerra* was open to all, and also because only a quarter of all annual promotions were the result of selection, but they spoke to the hostility felt toward the staff not only by the army at large but also by politicians with a presumed interest in its welfare.[25] In 1913, by which time the law on accelerated promotion had been modified six times in seventeen years, recruitment to the General Staff was extended from the

[22] Nicola Marselli, *La vita del reggimento* (Rome, 1903), pp. 15, 20.

[23] De Bono, *Nell'esercito nostro*, pp. 24, 96.

[24] "Il Corpo di Stato Maggiore ed I nuovi organici," *L'esercito italiano* 8 January 1882; "Le promozione nel Corpo di Stato maggiore," *L'esercito italiano* 20 December 1882; "L'avanzamento nell'esercito," *L'esercito italiano* 10 April 1885.

[25] "La relazione Compans," *L'Italia militare e marina* 14–15 June 1904; "L'onesta' dei fini e dei mezzi," *L'esercito italiano* 15 June 1904.

single grade of captain to higher officers of all combat arms who could pass set tests. Tradition died hard: on the eve of the war the old guard in the Senate, often an obstacle to reform and modernization, opposed "scholastic tests" to establish fitness for promotion by selection which the war minister wanted used more and not less frequently.[26]

Intelligence, Information, and the Planning Process

The formal establishment of an intelligence office – *Ufficio I* – at the General Staff did not occur until 1900. It was initially small, employing only four officers and enjoying a budget of 50,000 lire from which extraneous costs such as the expenses of trips made by the chief of the General Staff were deducted; six years after its foundation it employed only three agents operating outside Italy. If its wartime chief is to be believed, it did not even think of code-breaking and had to do a great deal of hard work after 1915 to catch up with both its enemies and its allies. Under Colonel Silvio Negri, its chief from 1905 until 1912, it made some strides forward, and with Pollio's support after he became chief of the General Staff, it gained more money and more staff. It also began to make use of irredentists from the Trentino as agents, a dimension of its work that dramatically expanded once the war began and one which produced some very effective results, particularly in counterespionage work. The department would appear to have languished somewhat under Negri's successor, Colonel Rossolino Poggi, in the face of the Finance Ministry's refusal to increase the section's budget to 200,000 lire.[27]

The General Staff gathered information by sending present and past members on unofficial tours to survey areas that were likely to become important in the event of war. Eugenio de Rossi, for example, was sent to explore the railway capacity in Galicia in 1907 and reported on the Chambéry-Grenoble-Briançon region in 1912.[28] While such activity produced important geographical intelligence, the main instruments for gathering more general information about the postures, attitudes and intentions of both actual allies and likely enemies were the military attachés. During Saletta's tenure of office (1896–1908), and as a result of the

[26] De Chaurand, *Come l'esercito nostro*, pp. 134, 151–2, 248; "La discussione della legge di avanzamento in Senato," *L'esercito italiano* 8 March 1913.
[27] De Rossi, *La vita di un ufficiale*, pp. 180–1; Odoardo Marchetti, *Il Servizio Informazioni dell'Esercito Italiano nella Grande Guerra* (Rome, 1937).
[28] De Rossi, *La vita di un ufficiale*, pp. 196, 226–7.

narrowly technical perspective which he took in the aftermath of the debacle at Adua and the consequent sensitivity of governmental relations with the chief of staff, attachés were kept on a fairly tight leash. In 1895, as deputy chief of staff, Saletta reproved Alberto Pollio, then military attaché at Vienna and subsequently his successor as chief of staff, for being too concerned with politico-military matters and told him to stick to his last.[29] When he in turn became chief of staff, and in circumstances in which international tensions were considerably greater, Pollio took a somewhat different line and sought politico-military information on such questions as whether Austro-German friendship (*intimita*) was diminishing or increasing.[30]

In 1889, the military attaché in Paris, Lieutenant-Colonel Massone, began sending back ever-increasing amounts of top secret information relating to experiments with new guns and explosives (melinite), plans of the fortifications at Nice, Toul, Albertville, and Besançon, appreciations of Italian defenses and the 1890 mobilization instructions for the French army. Some material was copied by Massone with the assistance of the naval attaché, and some was sent on to Rome for direct scrutiny and return by secret courier. The flow significantly diminished after 1892 and had more or less dried up by 1895. The material was almost certainly provided by Major Marie-Charles-Ferdinand Esterházy, for whose activities Captain Alfred Dreyfus was initially convicted.

In 1896, for reasons which are presently unknown but which may have had something to do with the Dreyfus case, the General Staff took the decision to renounce using "unsafe sources" and to rely even more heavily on the military attachés to pass on information.[31] The obligation not to make use of spies, either actively by recruiting them or passively by responding to offers of their services, was apparently formally incorporated into the regulations under which military attachés operated in 1906. Thereafter, the attachés seem to have stuck to the rules, reporting at intervals on their own observance of their instructions and on their practice of the self-denying ordinance in rejecting any overtures no matter how tempting, though on occasion Rome thought it desirable to remind them of their obligations. When the spy scandal of the Austro-Hungarian Colonel Alfred Redl (discussed by Günther Kronenbitter in his chapter) came to light in 1913, Lieutenant-Colonel Alberico Albricci, the military

29 AUSSME. N. 35, Saletta to Pollio, 30 March 1895. Rep. G-29 racc. 9.
30 AUSSME. Pollio to Barattieri, 3 November 1912. Rep. G-29 racc. 57.
31 AUSSME. N.23, Primerano to Mocenni, 6 January 1896. Rep. G-29 racc. 5.

attaché in Vienna at that time, assured Rome that he had had nothing whatever to do with the case.[32]

In practice, military attachés serving in the major European capitals worked together in groups reflecting the diplomatic alliances and political friendships currently in operation and these groups frequently shared information and exchanged assessments. This practice often put the Italian military attaché at a disadvantage. In Vienna, Captain Alessandro di San Marzano found himself isolated since Italy was not a member of the Triple Entente and the German military attaché was very reserved. This meant that he was particularly reliant on the press, which was censored; on the War Ministry, which responded slowly and incompletely to his requests for information; on his own observation, which was seriously inhibited by the frequent refusal of his requests to assist in garrison maneuvers, visit troops or be shown over military institutes; and on conversations with the community at large.[33] Similar circumstances made some postings almost impossible. The military attaché at St. Petersburg reported in 1912 that he was unable to get any information about the possible mobilization and deployment of the Russian army: "To the reserve of the military authorities is added the absolute silence of the press."[34]

Special instructions had to be provided for the military attaché to Turkey, who was warned that he would not get information in the usual way and advised that the British and Germans were the best-informed locally, that large firms, banks and railway companies were free with information, and that cooperation improved the farther away from Constantinople and the lower down the hierarchy one went. The attaché was also reminded, in line with Saletta's view of his functions, to touch on political questions only as far as was necessary in order to illuminate the military provisions being made "since nothing annoys our heads of mission overseas more than learning that the military attaché is talking or writing about politics."[35]

Despite all the problems they faced, the military attachés were generally able to provide a great deal of intelligence from open sources; even the Russian attaché was able to get some information from officialdom

[32] AUSSME. N. 142, Zaccone to Pollio, 4 September 1909; n.1055, Pollio to Albricci, 30 November 1909; n.15/18, Albricci to Barattieri, 19 January 1912; n.86, Albricci to Pollio, 15 April 1913. Rep. G-29 racc. 24, 14, 16, 17.

[33] AUSSME. N.97/138, Di San Marzano to Pollio, 6 November 1909. Rep. G-29 racc. 14.

[34] AUSSME. N.174, Abati to Pollio, 12 November 1912. Rep. G-29 racc. 86.

[35] AUSSME. Istruzioni per l'addetto militare a Costantinopoli, 24 March 1905. Rep. G-29 racc. 6.

as well as from the newspapers. The materials sent back to Rome ran the gamut from matters such as the trials and use of different pieces of equipment or weapons through military structures and systems of conscription to declaratory statements of general policy or intent. In the years immediately before 1914, in Berlin Colonel Luigi Calderari seems to have been able to get fairly ready access to Helmuth von Moltke (the Younger) and to his deputy, General Georg von Waldersee. He was also, from time to time, singled out by Wilhelm II as an avenue for the transmission of imperious *obiter dicta* about the state and tasks of the Triple Alliance. Albricci, in Vienna, was apparently equally able to have relatively uninhibited conversations with Baron Franz Conrad von Hötzendorf despite the Austrian chief of staff's deep and abiding distrust of the mass of the Italian populace.

War Plans, Staff Rides, and Maneuvers

When Cosenz became chief of the General Staff he found that nothing in the way of planning had been done by the ministers of war who had preceded him save for one, General Cesare Ricotti-Magnani (1870–76, 1884–87), who during his first period in office had decided the locations where the army corps would be concentrated – or so he claimed some years later. He laid down the parameters of planning by initiating studies of advanced defense to cover deployment, defensive and offensive war, and railway movement. In doing so, he consciously followed Moltke, intending in wartime only to provide general directives of offense or defense that would keep the various armies coordinated but making no attempt to direct actual operations.[36]

Italy's membership in the Triple Alliance brought with it both the obligation and, on Cosenz's part, the wish to help Germany as quickly as possible and draw off French forces. French fortification of the Alps and Swiss fortification of the St. Gotthard pass, coupled with German objections to breaking Swiss neutrality, left only one option – direct Italian cooperation on the Rhine. This in turn depended on being able to utilize Austrian railways to get there. Since the arrival of Italian troops in the main theater offered Austria the possibility of securing more German help in the East against Russia, there was more than a measure of self-interest behind its adherence to the military accord reached between the

[36] Istituto per la Storia del Risorgimento. Cosenz to Lionello De Benedetti, 12 September 1898. Carte Cosenz b. 108 n. 49/2.

three powers on 15 January 1888 under which Austria put three railway lines at Italy's disposal from the tenth day of mobilization in the event of a general war.

The details of Austrian assistance were embodied in a railway convention signed on 1 March 1888. A total of twenty-eight trains would run each day for twenty-three days (from the sixth to the twenty-ninth day of mobilization) on the three designated lines. Of the 577 locomotives that would be required, Italy could provide 100; Austria was not sure of being able to provide 234, and Germany would have to provide the remainder. Likewise, of the 10,000 railway wagons that would be needed, Italy could provide half, Austria one-fifth and Germany would provide the rest. At the end of that year, reversing Germany's earlier standpoint and sounding a note to which his successor would return, Count Alfred von Schlieffen suggested that if Austria remained neutral then Italy might breach Swiss neutrality to get to the communal battleground.[37]

The railway conventions were brought up to date in October 1891, when the number of trains went up to 623, the transit times were reduced by three days, and Germany undertook to produce 201 railway engines and 2,942 railway wagons. They were revised in 1893, in October 1896 when the Italians' arrival on the Rhine was speeded up by one day, and again in May 1898. During or shortly after the first meeting the probable Italian contingent was reduced from six to five of the total of twelve Italian army corps.

The likely role of Italian troops on the left wing of Schlieffen's "pincer" attack on France was indicated as early as November 1891, when the chief of the German General Staff sent Cosenz details of the forts of Épinal and Belfort. Two years later, the Italians received German studies of the invasion routes to Paris and Lyons.[38] However, the evidence that remains suggests that exchanges of concrete information did not go much beyond railway timetables. General Osio, designated chief of staff for the Italian III (Rhine) Army in the closing years of the nineteenth century, regretted "the scant information and nebulous directives" given him by the Germans while at the same time recognizing that "one could not ask for more from a people who considered us unreliable allies."[39] One project which was passed to Osio was for operations against Verdun:

[37] Massimo Mazzetti, *L'esercito italiano nella triplice alleanza* (Naples, 1974), pp. 37–8, 69–89, 110–3.

[38] AUSSME. N.132, Zuccari to Cosenz, 20 November 1891; n.18, Zuccari to Cosenz [?], 14 February 1893. Rep. G-29 racc. 50.

[39] De Rossi, *La vita di un ufficiale*, p. 138.

plans to transport three Italian army corps there were drawn up in the 1890s, for operations would have been entrusted to the Italians in the event of war. For their part, the Germans complained that although they gave the Italians whatever they wanted to know, especially regarding geographical data, they got nothing back.[40]

During the final years of Schlieffen's tenure signs of strain began to appear as the military intentions of both sides shifted. King Vittorio Emanuele III, newly installed on the throne after the assassination of his father in 1900, indicated in February 1901 that he was against the entire plan of committing five Italian corps on the Rhine because it weakened Italy too greatly. Although his attitude probably had something to do with his reluctance to cede royal command of a substantial part of his army to a foreign power, it also reflected his concern at the internal instability of Italy at that time. Schlieffen decided in 1902 that he could do without Italian aid on the Rhine. He did not tell the Italians so openly, but their military attaché found out. Saletta attended German maneuvers that year and reaffirmed that III Army would be sent to the Rhine, via Switzerland if it could not make use of Austrian railways.

The concept of breaking Swiss neutrality in order to link up with the Germans on the upper Rhine – known in Italian plans as the "second hypothesis" – had been incorporated into staff thinking since 1886, but only as a subject for study. It took a step towards practicality in 1898 when Saletta ordered the preparation of plans for Italian units to move through eastern Switzerland from the Valtelline via Chur to Lake Constance before turning west towards Zurich and Eglisau. These were duly communicated to Schlieffen and a general agreement was concluded in November 1900 according to which German sources would provide the munitions and supplies for the Italian III Army which, if it followed this route, would be out of reach of domestic sources of supply.[41]

In saying that III Army would make use of this option, if necessary, Saletta was disguising the problems it would present. The staff ride in 1900, which he had led, had examined the problem of violating Swiss neutrality in order to skirt north around the French fortifications at Albertville and Grenoble and had concluded that "an offensive moving from Italy towards Switzerland would encounter serious difficulties" because of the

40 Maria Gabriella Pasqualini, *Carte segrete dell'intelligence italiana (1861–1918)* (Rome, 2007), p. 164. AUSSME. Prudente to [?Saletta], 31 March 1899. Rep. G-29 racc. 53.

41 Mazzetti, *L'esercito italiano*, pp. 178–81.

shortness of the fighting season and the superiority of Swiss defenses over Italian ones, and that if Swiss neutrality were respected this "would eliminate the dangerous consequences of our inferiority."[42] Indications that Austria would be too preoccupied moving troops to Galicia to be able to get the Italian armies to the Rhine in time led to the Swiss option being closely examined in 1904–5. Despite German assurances that they would put pressure on German-speaking cantons to allow troop transits, the reexamination confirmed the findings of the staff ride and the idea was dropped.[43]

The "second hypothesis" formally disappeared from planning after 1907 but was not entirely abandoned: a staff ride in 1910 explored pushing four Italian corps north from the St. Gotthard pass to join German troops moving on the Rhine along the German-Swiss border via Lake Constance and Basel, but encountered immediate problems attacking the entrenched camps guarding the pass.[44] In 1911 a staff ride tested the scenario of a joint Austro-Swiss attack on Brescia, Bergamo, and Milan, and in February 1912 the War Ministry was sufficiently concerned about Switzerland's likely abandonment of neutrality in the event of war to raise the need for additional fortifications along that frontier as well as in the East and the West.[45]

At the end of 1903, Saletta again took pains to reassure the Germans that the Rhine army plan still stood despite Italy's tightening diplomatic links with France. The hostility to France that was one of the legacies of the *Risorgimento*, and that had been a marked feature of Crispi's reign as premier in the 1890s, had ended the previous year when, under the terms of the Prinetti-Barrère Accords, France had given Italy a free hand in Morocco in return for Italian neutrality if France was attacked or went to war as a result of provocation. How much the army knew about the secret accord is unclear, but clearly diplomatic change was in the air. Accordingly early in 1906, at a time when Italy's stance during the first Moroccan crisis was upsetting Germany, Saletta assured a skeptical Helmuth von Moltke in Berlin and the Austrian chief of staff,

42 AUSSME. Relazione del viaggio di stato maggiore dell'anno 1900, pp. 154, 155. Rep G-28 racc. 29.

43 AUSSME. Sunto di colloquio avvenuto il 29 maggio 1933=XI nei locali del Senato del Regno tra S.E. il generale Bonzani e S.E. il generale Zupelli. Rep. H-5 racc. 45/1933.

44 AUSSME. Relazione del viaggio di stato maggiore del 1907; Relazione sul viaggio di stato maggiore 1910, pp. 12–13, 120. Rep. G-28 racc. 37, 42.

45 ACS. Spingardi to Brusati, 17 February 1912, enclosing Spingardi to Pollio, 16 February 1912. Carte Brusati b.10/VI-4–36/321.

General Friedrich von Beck-Rzikowsky, that Italy intended to stick to her agreements.[46]

Italy's military options were in fact somewhat more restricted than Saletta made out, for its ability to participate in general operations on the Rhine was in fact hamstrung by its vulnerability to attack both from France and from Austria. As soon as he entered office, Saletta instituted a complete study of the Italian western and eastern frontiers through a series of staff rides which was begun in 1895 and concluded in 1898. His conclusion two years later was that only when "no longer restricted by military preoccupations with the powers of resistance of one frontier or the other" would Italian policy be "freed to follow the direction which most corresponds with the national interests."[47]

Interpreting Austria's policy and its military intentions was difficult: the 1904 Austrian staff ride along the Isonzo frontier and the coasts of Istria and Dalmatia had evidently threatening undertones, as did the testing of mountain equipment in the Tyrol during maneuvers the following year. However, the Italian military attaché in Vienna believed that the Austrians were preparing to intervene in the Balkans in order to open up the road to Salonika and thought that the measures being taken on the Italian frontier were only of secondary importance.[48] Saletta's preoccupation with the eastern frontier, manifested in the scenarios of the staff maneuvers of 1903, 1904, and 1905 which examined how well the defenses could stand up to an Austrian attack along the arc from the eastern Veneto to the western Tyrol, predated the appointment of Conrad as chief of the Austrian General Staff in November 1906. His arrival in office, together with the extensive and threatening Austrian maneuvers in Carinthia in September 1907, focused Italian attentions on her eastern frontier – where, according to one report, there was only one armored fort along the entire 600 kilometers of frontier. For the remainder of his period in office, Saletta was engaged in preparing mobilization plans to defend the eastern frontier, a task which also preoccupied Alberto Pollio when he took over as fourth chief of the Italian General Staff in June 1908.[49]

46 Mazzetti, *L'esercito italiano*, pp. 197–8, 202, 212–3.

47 AUSSME. Relazione del viaggio di stato maggiore dell'anno 1900, p. 166. Rep. G-28 racc. 29.

48 AUSSME. N.83, Del Mastro to Saletta, 29 April 1905; n.92, Del Mastro to Saletta, 15 May 1905; n.13, Del Mastro to Saletta, 17 January 1905. Rep. G-29 racc. 13.

49 Mazzetti, *L'esercito italiano*, pp. 229–30, 235–6; Adriano Alberti, *L'opera di S.E. il generale Pollio e l'esercito* (Rome, 1923), pp. 9, 19.

Pollio dealt initially with the deployment plans for a war against Austria, for which he moved the Italian front back from the Piave to the Tagliamento River. He also received intelligence that his two partners might be in the course of concluding a military convention for war against Italy.[50] Nevertheless, Pollio remained interested at least in the technical aspects of the Triple Alliance military convention, seeking more trains on the two available railway lines in order to cut the time taken before the Italians arrived on the Austrian frontier from twenty-three days to twenty-one days in September 1908 and updating the railway plan again in July 1911. But then, Italy's war in Libya which began in September 1911 absorbed much of his time and energy, as well as taking large numbers of troops from metropolitan Italy and using up extensive stocks of war material.

It was not, however, the consequences of the African venture that prompted Pollio to send Colonel Vittorio Zupelli to Berlin in advance of the renewal of the Triple Alliance on 5 December 1912 to say that III Army could no longer be sent to the Rhine. Nor was it a wish to keep Italian troops under Italian command and the consequent revival of two old projects for a crossing of the western Alps and a landing on the Provence coast during 1912.[51] The initiative was in fact taken by Moltke, who suggested to the Italian military attaché on 22 November that the two general staffs should communicate "given the imminent renewal of the Triple Alliance and what could arise from the current political situation." The moment, he felt, was "most serious" (*sehr ernst*). If Russia intervened in an Austro-Serbian conflict on the latter's side, this would be the *casus foederis* which would cause Germany to take the field and therefore in all probability trigger French mobilization.[52] Moltke wanted to know how far Germany could count on Italian help. As far as the Rhine army was concerned, he gave the strong impression that he would not insist on having it if Italy was thinking of using it in other operations.[53]

Pollio had wanted to talk to the Germans in October about the serious international situation, but had been reined in by General Paolo

[50] AUSSME. N.97/138, San Marzano to Pollio, 6 September 1909. Rep. G-29 racc. 14.

[51] *Contra* Alberti, *Pollio*, p. 20; Alberti, *Il generale Falkenhayn. Le relazioni tra I capi di S.M. della Triplice* (Rome, 1934), p. 67; Mazzetti, *L'esercito italiano*, pp. 258–9.

[52] AUSSME. Calderari to Pollio, 22 November 1912. Rep. H-5 racc. 45/'1912'.

[53] AUSSME. T.1539, Calderari to Pollio, 25 November 1912; T.204, Calderari to Pollio, 25 November 1912; Calderari to Pollio, 26 November 1912. Rep. H-5 racc. 45/1912. Calderari interpreted the sense of Moltke's remarks as being that it was desirable for Italy to immobilize two French army corps.

Spingardi, the war minister, who did not want to provoke a decision which might rescind the dispatch to the Rhine not only of five army corps but also of the two cavalry divisions scheduled to accompany them.[54] Pollio now told Spingardi that in view of the "abnormal circumstances" in which the army presently found itself, he did not think that Italy could maintain its agreements in their entirety "without grave prejudice to us." Accordingly, he wanted to know what the government's intentions were.[55] He was assured that it was not a matter of a new military convention on the basis of a likely renewal of the Triple Alliance treaty, but rather of "aid to Germany at the present time."

Pollio sent Colonel Zupelli to Rome to discuss the situation, but could give him no precise instructions because Moltke's proposals had been so vague. Zupelli was told simply to make no commitments without his, Pollio's, authorization. Zupelli went armed with a digest of the chief of staff's views of the strategic options, which were very limited: Italy could not count on the Austrians; crossing Switzerland would be "very difficult in war"; operating in the mountains on the French frontier would be difficult because of French fortifications and the need for siege artillery; and a war with France would expose Italy's coasts and its communications with Libya to attack. In the latter event, the only thing Italy could do would be to array part of its army along its northwest frontier and start a mountain war.[56]

Told that Italy could no longer send III Army to the Rhine, Moltke pointed out that it would be to Rome's advantage to do so and that it would be more effective to act against French troops behind the Alpes Maritimes aided by the vigorous German offensive against France rather than attempt a long and difficult crossing of the mountains.[57] In the course of the discussions in early December, Moltke told the Italian staff officers that if Serbia went to war with Austria, then Vienna could count on Germany.

54 ACS. Spingardi to Brusati, 26 October 1912. Carte Brusati b. 10/VI-5–37/361.

55 AUSSME. N.255, Pollio to Spingardi, 25 November 1912. Rep. H-5 racc. 45/1912. This letter contains the sentence: "[I]n view of the pacts by which our army is linked to the German army, and in case the *casus foederis* were to be verified for us too, I think it opportune that we send a delegate to Berlin." The phraseology raises a small question mark over the commonly asserted belief – to which Pollio himself contributed – that the chief of the Italian General Staff was kept in ignorance of the nature of Italy's diplomatic commitments.

56 AUSSME. Istruzioni pel colonello Zupelli, 27 November 1911. Rep. H-5 racc. 45/1912.

57 AUSSME. T., Zupelli to Pollio, 4 December 1912. Rep. H-5 racc. 45/1912.

Pollio found himself unable to give his German opposite the answers he needed about what Italy would do in particular circumstances because, as he complained to the war minister, he was absolutely ignorant of the clauses of the Triple Alliance that linked Italy to Germany. He only knew that the Triple Alliance was being renewed because the emperor had told Calderari.[58] Pollio absolutely excluded sending an army to Germany "for now" and asked the government two questions: if the *casus foederis* was verified, would Italy mobilize simultaneously with Germany, and would the government order a general mobilization or not? A letter that spoke volumes about the nature of civil-military relations in Giolittian Italy concluded with a request from the chief of the General Staff to be kept abreast of the international political situation as stipulated by the royal decree of March 1908, and the statement that he would take the king's orders as to the actual employment of the troops.[59] In reversing the decisions that had held since 1888, Pollio had two operational considerations at the forefront of his mind: the war in Libya and the need to prepare for "complications" in the Balkans, for which III Army must be held as a strategic reserve.[60]

Before his interview with Zupelli, Moltke had been palpably unsure of Italian collaboration; afterward, he was apparently more optimistic, remarking that a resolute Italian commitment would make Germany's military effort "a certain success."[61] When Conrad, newly reinstalled as chief of the Austrian General Staff, was told of the decision not to send III Army to the Rhine, he appeared "surprised and affected" by the news and asked whether all Italian forces would be deployed in the western Alps and where the rest would go if they were not.[62] His interest in the location of the Italian strategic reserve was one of a number of signs that he doubted Italy's intentions, and he believed its location might betray Rome's intention not to commit itself to the letter of the Triple Alliance, but to act according to interests that were "not presently admissible." A general feeling was abroad in Vienna that Italy acted "in bad faith."[63]

[58] AUSSME. N.219, Calderari to Pollio, 23 November 1912. Rep. G-29 racc. 57.

[59] AUSSME. N.2263, Pollio to Spingardi, 7 December 1912. Rep. H-5 racc. 45/1912. Unfortunately, Spingardi's reply does not survive in the archive. Pollio complained again two weeks later that he did not know the text of the treaty and whether Italy had undertaken to support Austria-Hungary in any particular circumstances: ACS, Pollio to Brusati, 24 December 1912. Carte Brusati b.10/VII-2–42/442.

[60] AUSSME. T.275, Pollio to Albricci, 28 December 1912. Rep. H-5 racc. 49/4.

[61] Michael J. Palumbo, "German-Italian Military Relations on the Eve of World War I," *Central European History* XII (1979), p. 348.

[62] AUSSME. N.145/352, Albricci to Pollio, 18 December 1912. Rep. H-5 racc. 49/4.

[63] AUSSME. Albricci to Pollio, 3 January 1913. Rep. H-5 racc. 45/1913.

During the December conversations leading to the abrogation of the military convention of 1888, Pollio evidently betrayed concerns about Swiss neutrality and about Italy's situation in a general war when Germany turned to face Russia, as he knew it would. Moltke's reassurances that Switzerland would remain neutral and that when the turn to the East came Italy would not have to confront France alone as the German General Staff foresaw putting German army corps under Italian command, contributed to his willingness to reconsider sending Italian forces to the Rhine even at the moment when he was making it plain that strategic circumstances presently prevented him from doing so.[64]

Faced with Italy's *fait accompli*, Moltke took up a suggestion made by Pollio during the staff conversations for a Triple Alliance naval convention, which would both improve the chances for Italian landings on the southern coast of France and prevent France from sending reinforcements from North Africa which would amount to almost two army corps.[65] He also sent General von Waldersee to Rome in January 1913 to discuss "unity of action" – a phrase that puzzled Pollio, given the space and time that separated the two theaters in which the Italian and German armies were now planning to fight.[66] Cavalry was the only arm of which Pollio had spare capacity, and Waldersee was given to understand that some would be sent to the Rhine. According to German accounts, Pollio also promised to make a full army available for service on the Rhine after he got royal and political approval.

Pollio's willingness in principle to contribute to the common struggle in which he was becoming an increasingly convinced believer was evident in the deployment plans drawn up by the General Staff in January 1913: while three corps crossed the Maritime Alps, one corps stood at Milan to watch the Ticino and five were held in reserve to hinder French landings or invasion, VII army corps stood at Padua ready to go to Germany.[67]

The outbreak of the second Balkan war on 3 February 1913, and the threat to Albania that developed in April as a consequence, temporarily displaced war planning among the Triple Alliance partners. This displacement produced an order for the secret mobilization of the Italian army and navy on 30 April and strong complaints from Pollio that he was not being kept up to date with political matters, that politicians did not

64 AUSSME. T.265, Zupelli to Pollio, 9 December 1912. Rep. H-5 racc. 45/1912.
65 AUSSME. Moltke to Pollio, 9 January 1913. Rep. H-5 racc. 25/1913.
66 AUSSME. Colloquio con S.E. il Capo di S.M. dell'esercito, 15 January 1913. Rep. H-5 racc. 45/1912.
67 Mazzetti, *L'esercito italiano*, pp. 290–1.

appreciate the time needed for mobilization, and particularly that he was "in the dark" about the diplomatic agreements with Austria.[68]

Meanwhile, however, pressures were building up to reverse the decision Pollio had taken the previous December. In May 1913, General Luigi Cadorna, commander-designate of II Army, bombarded Pollio, Spingardi and the king's chief military aide-de-camp, General Ugo Brusati, with memoranda castigating the French Alps as a secondary military theater where action would be indecisive and against which Italy would break her head while achieving nothing.[69] At about this time, or more probably a little later, an outline plan for Italian landings on the coast of Provence demonstrated that military operations there could not begin less than four months after the opening of general hostilities.[70]

In August 1913, Pollio went to Germany to attend the annual summer maneuvers. While there he made a good impression on Moltke, who found him "most agreeable," and on Conrad, who "had the feeling that he was a sure supporter of faithfulness to the alliance."[71] In meetings with Waldersee at the beginning of September, Pollio learned that the balance of forces was such that Austria, facing Serbia and Greece in the Balkans, would need German reinforcements to strengthen it against Russia, that these reinforcements would be drawn from the left wing of the German armies scheduled to attack France, and that therefore any Italian reinforcements in that sector would be of great if not decisive importance. Pollio agreed to try to persuade the king to send two army corps comprising five divisions to operate on the German left wing against France, and indicated that he was willing to consider sending the two Italian cavalry divisions to Silesia to operate against Russia, though for technical reasons he felt they would be better employed against France.[72] After the meeting Conrad, evidently somewhat reassured, hoped that Italian troops would be included in the Triple Alliance's military calculations for 1914–15.

On 10 October 1913, the three chiefs of staff met at Salzbrunn in Silesia under the supervisory eye of Wilhelm II. After Conrad had explained

[68] ACS. Pollio to Brusati, 1 May 1913. Carte Brusati b. 10/VII-3–43/481.

[69] ACS. Cadorna to Brusati, 26 May 1913; Cadorna to Pollio, 26 May 1913; Cadorna to Spingardi, 26 May 1913. Carte Brusati b. 10/VII-I-41/408.

[70] Paul Halpern, *The Mediterranean Naval Situation 1912–1914* (Cambridge MA, 1971), pp. 270–2. The memorandum is undated and its timing is disputed; Halpern suggests 1913-early 1914, but Mazzetti favors late spring-summer 1913. Mazzetti, *L'esercito italiano*, pp. 365–6 fn. 26. See below, fn. 73.

[71] Alberti, *Il generale Falkenhayn*, pp. 70, 71.

[72] Mazzetti, *L'esercito italiano*, pp. 350–4; Palumbo, "German-Italian Military Relations," p. 350.

the advantages to Austria of replacing German troops in the West and sending them to the East where they would facilitate an offensive against Russia, Pollio rehearsed the military arithmetic behind his decision to advise the annulment of the military convention regarding III Army. With three corps in Libya and the Aegean, one on stand-by for possible use in Albania, and five on the Rhine, only three corps would remain to fight a war on the Alpine front and defend the entire Italian peninsula. Taking the line initially set out by Moltke, he declared himself "persuaded that in a war the Triple Alliance must act as a single state, dealing [as we shall be] with a question of [our very] existence, since the war will be terrible," and confirmed that he was prepared to put two divisions of cavalry at Germany's disposition provided that he got official sanction.

He had always intended to return to the matter of III Army when the situation permitted, he told his audience, and he listed the practical issues that led him to want to do so: the operational season in the Alps was short; not many troops could be employed there for logistical and tactical reasons; and he needed to find useful work for at least one-third of Italy's mobile forces because France would not be able to employ large forces for landings on the Italian coastline. A landing by Italian troops at the mouth of the Rhône was "in the course of study," but because it depended on first securing a complete victory at sea, it was "risky." After interventions by both Moltke and the kaiser, it was agreed that the Italian cavalry would be better located on the Rhine. Pollio concluded by confirming that he was studying the employment there of two army corps.[73] In the course of the meeting, the Italian chief of staff was given a clear picture of how Germany planned to act in the West in the event of war. France was protected by a formidable belt of fortifications along her eastern frontier "but the German General Staff hope to stave in this belt, to pass beyond it and to carry out a short but terrible war."[74]

In December 1913, Waldersee was sent to Rome to press the case for Italian troops on the Rhine, but though Pollio hoped soon to be able to put three army corps at Moltke's disposal, he could make no promises. However, when Italian delegates met in Vienna at the end of January 1914 to arrange the transport of the two cavalry divisions, which the Germans

73 AUSSME. N.103, Pollio to Spingardi, 12 October 1913. Rep. H-5 racc. 45/1913.

74 AUSSME. N.104, Pollio to Spingardi, 19 October 1913. Rep. H-5 racc. 45/1913. This letter also confirms that studies for the operation to land in Provence were still in progress and that Pollio was unconvinced by them: "the more they progress, the less it seems to me the likelihood of a good outcome."

wished to locate either at Strassburg or at Freiburg and Colmar, Pollio was happy to have the movement of three army corps studied as well.[75]

During the next ten days, as a result of a process that is still obscure, the Italian government reversed the decision of December 1912 and secured the king's assent to send three Italian army corps to the Rhine. Moltke was told of the decision on 11 February and Conrad on 16 February. Berlin moved quickly: on 10–11 March General Luigi Zuccari, newly designated commander of III Army, attended a meeting with Waldersee in Germany at which he was informed that since the mobilization regulations for 1914 had already been prepared, no account could be taken in them of the three Italian corps, but that in 1915 they would be expected by the end of the third week of mobilization and would be allotted the task either of launching a diversionary attack on French frontier fortifications or of counterattacking the French in Lorraine. On 10 April a transport agreement was signed in Vienna which would shuttle three Italian army corps to Germany on 541 trains operating along three lines, concluding by the twenty-sixth day of mobilization.[76]

In the course of discussions over the new military convention which would come into operation in the event of the *casus foederis*, Moltke indicated that he now wished not to specify any fixed locality for the deployment of Italian forces since the military situation could be such as to require their employment elsewhere than on the Rhine. The draft convention for 1914–15 mentioned only the possible use of the three Italian corps in attacks on the *Sperrforts* of the Upper Moselle or the fortifications at Belfort and Épinal.[77] However, Moltke and Waldersee were now thinking of deploying the Italian contingent toward the Russian frontier where "because of the new European constellation the assistance that Germany can give the Austro-Hungarian army has become less appreciable."[78]

The issue of timing seems to have played a significant part in this literal volte-face; Moltke expected the decisive battle in the West to take place

[75] AUSSME. Fiastri to Pollio, 30 January 1914; T.42, Pollio to Fiastri, 31 January 1914. Rep. H-5 racc. 49/4. *Contra* Mazzetti, *L'esercito italiano*, p. 378.

[76] Mazzetti, *L'esercito italiano*, pp. 381, 385, 393–4. Since the mobilization year began on 1 April, the Germans could not take account of three Italian corps on the western front that year, *contra* Palumbo, "German-Italian Military Relations," pp. 351–2.

[77] AUSSME. Note alle osservazioni dello St. M. germanico circa le proposte dello St. M. italiano relativamente alla convenzione militare 1914 [n.d. ? 1934], allegato (b). Rep. H-5 racc. 45/1914.

[78] AUSSME. N.88, Calderari to Pollio, 23 May 1914. Rep. H-5 racc. 45/1914. The king had caught wind of this idea six weeks earlier – it is not clear how – and Pollio had to reassure Brusati on 9 April that the idea was for internal study only: Mazzetti, *L'esercito italiano*, pp. 394–5.

in the first three weeks of war and the Italian troops would only arrive on the Rhine during the fourth week. On the eve of the assassination of Archduke Franz Ferdinand, shortly before his own unexpectedly early death, Pollio recognized that the course of events once a war had started might justify deploying Italian troops against Russia and was willing to assume responsibility for such a deployment after war had been declared, though he was not presently willing either to treat with the Italian government over it or to have it inserted in the new military convention.[79]

Although the eleventh-hour change of heart in Rome meant that Italo-German military planning was at sixes and sevens on the eve of the war, relations between Rome and Vienna were growing tenser. Irredentist activity in Italy over the previous winter had reawakened scarcely dormant animosities, and on 1 June 1914 the Italian military attaché reported the growing strength of anti-Italian feeling among clerico-aristocratic circles in Vienna, within which Franz Ferdinand was at the fore. Italy was regarded as an untrustworthy ally, the treatment of Italians inside the monarchy was becoming intolerable (at Pola sailors and dockyard workers were reportedly forbidden to speak Italian or to marry Italians), and the state was persistently arming against Italy.[80]

On 24 June, seven days before his death, Pollio sent the War Ministry a memorandum pointing out that Austria had almost as many troops on the Italian as on the Russian frontier, that in 1914 she had spent almost one milliard lire on her land forces, and that Italy would only be safe from the Austrian threat when she was as strong as her neighbor.[81] As well as reflecting a traditional political anxiety about Italy's northeastern neighbor and an anxiety about the defensive capabilities of the eastern frontier which was by now well established in the planners' minds, this may well have been a calculated move in the ongoing contest for funding.

In May and early June 1914 Moltke and Waldersee, as well as Prussian War Minister Erich von Falkenhayn, expected Italian military support in the event of a war between the Triple Alliance and its enemies. Conrad had less faith in Italy, despite Moltke's efforts to persuade him that "the frank and loyal words of General Pollio may be a guarantee for [us] all," because of the anti-Austrian feelings nursed by a large part of the Italian population.[82] But diplomats felt less certain. After the assassination of Franz Ferdinand at Sarajevo, Prince Karl von Lichnowsky, the German

[79] AUSSME. T.149, Pollio to Zuccari, 16 June 1914. Rep. H-5 racc. 45/1914.
[80] AUSSME. Albricci to Pollio, 1 June 1914. Rep. H-5 racc. 49/4.
[81] Mazzetti, *L'esercito italiano*, pp. 409–12.
[82] AUSSME. N.107, Calderari to Pollio, 23 June 1914. Rep. G-29 racc. 58.

ambassador to Britain, warned that Italy could not be counted on, but was told that it had given the Triple Alliance assurances "of the friendliest kind."[83]

Pollio died on 1 July and Luigi Cadorna was named as his replacement on 10 July, formally taking up the post of chief of the General Staff seven days later. Pollio's death seemed to have made no difference to Italy's military posture, for Cadorna wrote to Moltke on 27 July fully accepting the agreements made by his predecessor and assuring the chief of the German General Staff about the bonds of comradeship that united Italy and Germany. Moltke's reply two days later speculated that very soon the two armies would be united in battle and working towards a common victory.

Consistent in his opposition to the idea of an Italian campaign in the western Alps, on 31 July Cadorna sought the king's agreement to increase the size of the Italian force to be sent to the Rhine in the event that the *casus foederis* was verified. After the war he said that he would have sent nine of Italy's twelve available army corps. On 2 August, Vittorio Emanuele approved his proposal. The next day Cadorna received a letter from Conrad informing him that his deceased predecessor, in what were on the Italian side hitherto unknown verbal communications, had said that in addition to sending forces to Germany he would make available troops for the direct support of Austria-Hungary. The Austrian chief of staff now asked what they were and when and where they would be available.[84]

The news that in the eyes of Italy's leadership the *casus foederis* did not exist, telegraphed to Wilhelm II by Vittorio Emanuele III on 3 August, shook Moltke; two days later he forecast that Italy's "crime" would be avenged in history. "God grant you victory now," he told Conrad, "so that later you will be able to settle accounts with these rascals."[85] Cadorna's immediate concern was that German or Austrian troops might decide to attack Italy and that Switzerland might not contest their passage across her territory. Given that Italy was now the only neutral avenue along which Switzerland, which only possessed one month's

[83] John C. G. Röhl, *Zwei Deutsche Fürsten zur Kriegsschuldfrage* (Düsseldorf, 1971), quo. Palumbo, "German-Italian Military Relations," p. 361.

[84] Conrad to Cadorna, 1 August 1914: Alberti, *Il generale Falkenhayn*, p. 82. Montanari afterwards recorded that when Pollio sent him to Vienna he suggested saying that sending Italian troops to Vienna was not to be excluded if there were excess troops left in Italy and no more could be sent to Germany: fn. 76 above.

[85] Moltke to Conrad, 5 August 1914, quo. Mazzetti, *L'esercito italiano*, p. 447.

supply of grain, could import replenishments, he wanted maximum pressure exerted to get a formal guarantee from her that Bern would oppose "forcibly" any attempt to violate its territory in order to invade Italy.[86]

Conclusions

Moltke had good reason to be both upset and angry, for his plans had been seriously undone. Had Italy stuck to her modified military agreement, then the combined effect of attacking the French across the Alps and preventing North African reinforcements arriving in the *metropole* could have made a difference of twenty divisions in the balance of Franco-German forces.[87] As it was, the "left hook" on the Rhine, which had been the feature of prewar planning for almost a decade and which would almost certainly have reappeared in 1915 had war not occurred in August 1914, had to be abandoned. So did the plan to send *Ersatz* divisions to the East to help Conrad, troops which were the more necessary because of the loss of twenty divisions as a result of Romania's declaration of neutrality, but which now had to be used in the West instead.

Thus, the Italians' disappearance from the military scene had major effects on both fronts. Indeed, it may provide the explanation for the puzzle as to why, having never fully practiced Schlieffen's 1905 war plan before August 1914, Moltke then abandoned his prewar design for war on the western front and instead adopted his predecessor's ambitious and essentially impractical idea, gambling on a sweep through Belgium to outmarch, outmanoeuvre and outfight French and British forces operating on interior lines.[88]

Had the war been delayed for a year, and had Italian troops appeared as planned on the Rhine to help encircle the French armies in Lorraine, their contribution might have had as decisive an effect as did their absence when war actually came, though the Italian General Staff's projections of what would happen after the guns began firing were as nebulous and uncertain as those of all the other general staffs about to face the test of war.

86 AUSSME. N.513, Cadorna to Ministry of War, 13 August 1914. Rep. H-5 racc. 45/1914.
87 Palumbo, "German-Italian Military Relations," p. 359.
88 See Terence Zuber, *Inventing the Schlieffen Plan: German War Planning 1871–1914* (Oxford, 2002), pp. 221–304.

8

Conclusions

Holger H. Herwig

Why did none of the "war plans" of 1914 succeed? Why did they all prove to be inadequate in some way or another? Why did the much-anticipated Armageddon not occur either in Galicia or in northern France? And why were the troops not home by Christmas after the "short, cleansing thunderstorm" predicted by both popular and military writers? The authors of the foregoing essays have addressed these questions through evaluations of the individual war plans of the great powers in 1914. War plans, in the inimitable words of historian Dennis Showalter, were "the nineteenth-century intelligence equivalent of the medieval knight's Holy Grail."[1]

"The history of the campaign of 1914 is nothing else but the story of the consequences of the strategical errors of the War Plan."[2] With these critical words, General N. N. Golovin sought to explain the Russian debacle of 1914. There obviously was no doubt in his mind that war plans existed – at least in Russia – on the eve of the July Crisis 1914 and that the opening moves of World War I were executed by the great captains on the basis of the plans at hand. In Germany, Staff Chief General Wilhelm Groener after the war spoke of the "symphony" of the Schlieffen Plan and of the "conductor" (Helmuth von Moltke the Younger) who "bungled" its execution.

[1] Dennis E. Showalter, "Intelligence on the Eve of Transformation," in Walter T. Hitchcock, ed., *The Intelligence Revolution: A Historical Perspective* (Washington, DC, 1991), p. 18.

[2] N. N. Golovine [*sic*], *The Russian Campaign of 1914: The Beginning of the War and Operations in East Prussia* (Fort Leavenworth, KS, 1931), p. 73.

For the historian of today, the task is not so simple nor the path so straightforward. The very term "war plan" conjures up images of military planners in secret drafting precise instruments for the deployment and operations of million-man conscript armies, plans which those in power in 1914 merely had to take off the shelves, dust off, and execute. The word "plan," defined by *Webster's* as "a method of achieving an end," suggests something structured, institutionalized, formal. Yet, the terms "war plans" and "war planning" are not to be found in the classic works of some of the greatest military thinkers such as Sun Tzu and Carl von Clausewitz.[3] Nor are they indexed in two of the more recent major treatises on strategic planning, *Makers of Modern Strategy* and *The Making of Strategy*.[4] Not even the United States Joint Chiefs of Staff tried to define the concepts in their *Official Dictionary*; the closest they came to a definition was under the heading "planning factor."[5] A most recent, deceptively clever reexamination of the boldest of the 1914 designs goes so far as to deny what every German general knew to be the case in 1914 – namely, that there existed a "Schlieffen Plan" and that the German General Staff executed its overall design, rightly or wrongly.[6] Little wonder, then, that when the authors of the case studies in this volume met to define their individual contributions at the Mershon Center of The Ohio State University in March 2005, each registered concerns about using the term "war plans."

And there are good reasons for concern. For, in none of the cases studied was there a sort of "doomsday" mechanism for automatic war as made famous in Stanley Kubrick's classic film *Dr. Strangelove*. Instead, the staffs in at least the continental nations constantly devised and revised what was a system of dynamic, fluid, ever-evolving war plans. They reacted to ongoing diplomatic, military, and domestic affairs. They

[3] Sun-tzu, *The Art of War* ed. Ralph Sawyer (Boulder, CO, 1994), p. 374; Carl von Clausewitz, *On War* eds. Michael Howard and Peter Paret (Princeton, 1976), p. 722.

[4] Peter Paret, ed., *Makers of Modern Strategy from Machiavelli to the Nuclear Age* (Princeton, 1986), p. 940; Williamson Murray, MacGregor Knox, and Alvin Bernstein, eds., *The Making of Strategy: Rulers, States, and War* (Cambridge, 1994), p. 679.

[5] Joint Chiefs of Staff, *The Official Dictionary of Military Terms* (Cambridge, MA, 1988), p. 274.

[6] Terence Zuber, *Inventing the Schlieffen Plan: German War Planning, 1871–1914* (Oxford, 2002), p. 299. Still, as the subtitle suggests, the author does acknowledge the existence of German "war planning" before 1914. Zuber's argument has been critically undermined by the research of Gerhard P. Groß, "There was a Schlieffen Plan," in Hans Ehlert, Michael Epkenhans, and Gerhard P. Groß, eds., *Der Schlieffenplan. Analysen und Dokumente* (Paderborn, 2006), pp. 117–60.

measured their labors against past experiences, current realities, and future prognostications. This, after all, was their proper professional function.

Thus, as the numerical iterations indicate, France had formulated no fewer than seventeen mobilization plans against Germany by 1914. Russia had penned no fewer than nineteen war plans against just Austria-Hungary and Germany on its western borders, not to mention those plans devised against Japan in the East or the Ottoman Empire in the South. At Vienna, General Franz Baron Conrad von Hötzendorf yearly updated his war plans. In the seven years before 1914, he had the 700 members of his staff draft plans for war against Italy and Russia (1907); Russia, Serbia and Italy (1908); Serbia and Montenegro (1909); Italy (1910); Italy, Serbia, and Montenegro (1911); Russia and Serbia (1912); and Albania, Montenegro, Russia, and Serbia (1913).[7] Additionally, Conrad drafted various defensive plans in case of an attack by Romania. Each year Conrad dutifully presented his plans to Kaiser Franz Joseph for approval. Even Germany's junior service was busily engaged in war planning: the Imperial Navy's operations plans targeted Belgium, Britain, Holland, France, Russia, and the United States.

But these plans were unofficial, "contingency plans" in the purest sense of the term. All required formal approval for implementation. None of the planners occupied responsible governmental or constitutional positions. Most served at the pleasure of the ruling "absolute and unquestionable" monarch, in whose hands the "war powers" rested. The chief of the German General Staff was but "the first military adviser" to the kaiser. In France, the "war powers" formally lay with the president, but required the countersignature of a minister and consent of the Senate and the Chamber of Deputies. In Britain the "war powers" rested with the prime minister and his Cabinet, and the (unstated) consent of Parliament.[8] Had the Great War broken out over any of the previous international crises – Morocco in 1905, Bosnia-Herzegovina in 1908, Libya in 1911, Morocco again in 1911, or the Balkans in 1912 and 1913 – the "plan of the moment" would have served as the basis for armed conflagration. The plans we deal with

[7] Holger H. Herwig, *The First World War: Germany and Austria-Hungary 1914–1918* (London, 1997), p. 9. Austro-Hungarian military planning before 1914 has been analyzed by Günther Kronenbitter, *"Krieg im Frieden." Die Führung der k.u.k. Armee und die Großmachtpolitik Österreich-Ungarns 1906–1914* (Munich, 2003).

[8] See the various chapters in Richard F. Hamilton and Holger H. Herwig, eds., *The Origins of World War I* (Cambridge, 2003).

thus gained importance, and indeed legitimacy, primarily because they were implemented in August 1914.

Before examining the individual war plans of the six major European powers, a few general observations are in order. First, virtually all the powers viewed war not as an immoral or as a desperate act but rather, in Clausewitz's oft-cited words, as the "extension of politics by other means." Germany and Italy had become national states through war. Russia had expanded its territory into Central Asia through war. With the exception of Austria-Hungary, each of the major powers of 1914 had conducted colonial wars. Britain under Queen Victoria was perhaps the most glaring example, having, according to historian Byron Farwell, mounted military campaigns during every year of her long reign (1837–1901).[9]

Second, as the authors of the foregoing essays show, most military planners were beset by a certain common *mentalité*. They had accepted the basic tenets that states rose and fell according to their ability to exercise and to project power, and that interstate rivalries could be resolved only by the recourse to arms. Put differently, on the eve of World War I the European powers were convinced of what historian Wolfgang J. Mommsen called the "topos of inevitable war."[10] To be sure, this assertion recently has been challenged by historian Holger Afflerbach, who argues that both Berlin and Vienna were dominated by "the fundamental desire" on the parts of their political and diplomatic decision makers "to preserve the European peace."[11] But the countless prewar utterances of those leaders concerning the "inevitability" of war in Europe – be it for economic, geopolitical, imperial, racial, or strategic reasons – suggests that Mommsen's claim remains valid. Moreover, the hope nurtured by many political advisors – that no power would risk escalating minor conflicts into a continental war, that each would "bluff" the adversary up the escalatory ladder stopping just short of war in favor of diplomatic resolution – proved to be another victim of July 1914.[12]

[9] Byron Farwell, *Queen Victoria's Little Wars* (New York, 1972), p. 1.

[10] Wolfgang J. Mommsen, "The Topos of Inevitable War in Germany in the Decade before 1914," in Volker R. Berghahn and Martin Kitchen, eds., *Germany in the Age of Total War* (London, 1981), pp. 23–45.

[11] Holger Afflerbach, *Der Dreibund. Europäische Großmacht- und Allianzpolitik vor dem Ersten Weltkrieg* (Vienna, 2002), pp. 24–33, 813–16, 823–25, 873.

[12] See J. J. Ruedorffer, *Grundzüge der Weltpolitik der Gegenwart* (Suttgart and Berlin, 1914), p. 219. "Ruedorffer" was a pseudonym used by Kurt Riezler, Chancellor von Bethmann Hollweg's political counselor at Berlin.

Third, as the Franco-Prussian War (1870–1) seemed to have revealed, the advantage in military campaigns rested with the offensive.[13] Modern fire power rendered stone fortresses redundant. It cleared the battlefield of encumbrances to aid infantry advances and cavalry sweeps. This "cult of the offensive," as political scientist Steven Van Evera has described it, dominated planners throughout the late nineteenth century.[14]

Fourth, military elites, at least publicly, nurtured what historian Lancelot Farrar has called the "short-war illusion."[15] Modern industrial societies with their tightly interwoven financial and industrial institutions and their delicate domestic social balance, the argument ran, could not sustain prolonged coalition wars for fear of domestic chaos and revolution – what Alfred von Schlieffen called the dreaded "red spectre."[16] Hence, military planners in all the major capitals started with the "given" that wars had to be short and decisive.

Fifth, European armies had grown enormously in the preceding centuries. Initially, these forces had been small. In 1630, Gustavus Adolphus had transported the entire Swedish army of 16,000 across the Baltic Sea to Peenemünde, Germany; in 1708 his successor, Charles XII, had crossed the Vistula River with a force of 20,000 foot soldiers. But just fifty years later, Frederick II of Prussia launched the Seven Years' War (1756–63) with armies of 150,000 men. In 1812, Napoleon I invaded Russia with a *Grande Armée* of as many as 690,000 soldiers, including 450,000 French. In July 1870 the Elder Moltke mobilized 380,000 Prussian and other German soldiers against France. And the best estimates for the U.S. Civil War suggest that about 2 million individuals served with the Federals and 750,000 with the Confederates. These vast forces, usually divided into army corps, armies, and later army groups demanded professional staffs for their mobilization, deployment, and resupply.

[13] See Geoffrey Wawro, *The Franco-Prussian War: The German Conquest of France in 1870–1871* (Cambridge, 2003).

[14] Stephen Van Evera, "The Cult of the Offensive and the Origins of the First World War," *International Security* 9 (1984), pp. 58–107.

[15] Lancelot L. Farrar, *The Short-War Illusion: German Policy, Strategy & Domestic Affairs, August–December 1914* (Santa Barbara, 1973). For a more recent assessment, see Holger H. Herwig, "Germany and the 'Short-War' Illusion: Toward a New Interpretation?," *The Journal of Military History* 66 (2002), pp. 681–93.

[16] Alfred von Schlieffen, "Der Krieg in der Gegenwart," *Deutsche Revue* (January 1909), pp. 13–24. Much of this had been suggested by Jan Gotlib Bloch, *The Future of War in its Technical, Economic, and Political Relations: Is War now Possible?* (New York, 1899).

A brief glance at the internal workings of the Prussian General Staff will specify these generalizations. The *Generalstab* grew from a mere 15 officers in 1870 to 650 in 1914. Six major sections handled its workload: Central Section controlled the flow of paperwork and monitored work completion; First and Third Sections dealt with intelligence and analyzing foreign armies; Second Section was responsible for mobilization and hence railroad schedules; Fourth Section directed staff rides and supervised the War Academy; and Fifth Section researched military history.[17] Competition for staff billets was fierce; promotion was based on brutal annual fitness reports. In short, warfare by the end of the nineteenth century had become a highly structured and "scientific" undertaking.

The transition from peace to war in Imperial Germany, for example, was managed through seven distinct phases of mobilization. In 1914 about 200,000 telegraph employees and 100,000 telephone operators communicated the details of the German mobilization to 106 infantry brigades; 23 railroad directorates commanded 30,000 locomotives as well as 65,000 passenger and 800,000 freight cars. Stage five of the so-called Travel Plan was the most critical, as it depended on the equal and constant speed and intervals of troop trains to move roughly 2.1 million men and 600,000 horses on the same stretches of tracks.

Time translated into space and blood in modern war planning. French Chief of Staff Joseph Joffre, for example, warned President Raymond Poincaré during the July Crisis that "Every delay of twenty-four hours in calling up reservists and sending the telegram for *couverture* [frontier covering force] means . . . the initial abandonment of between fifteen and twenty kilometers of territory for every day of delay."[18] Armies of communications and transportation experts labored tirelessly to shave days and even hours off complex mobilization timetables. A major last-minute shift in these schedules, as the Austro-Hungarian army discovered in late July and early August 1914, could bring chaos – and in that case required the immediate revision of no fewer than 84 boxes of detailed instructions.[19]

17 Arden Bucholz, *Moltke, Schlieffen and Prussian War Planning* (New York and Oxford, 1991), pp. 300–02.

18 Joseph Joffre, *Mémoires du maréchal Joffre (1910–1917)* (2 vols., Paris, 1932), vol. 1, p. 222; Raymond Poincaré, *Comment fut déclarée la guerre de 1914* (Paris, 1939), pp. 119–20.

19 Norman Stone, "Die Mobilmachung der österreichisch-ungarischen Armee 1914," *Militärgeschichtliche Mitteilungen* 16 (1974), p. 91.

The scope of war planning simply had become staggering in an age of industrial warfare, one in which it was generally expected (or feared) that reserves would be deployed from the start. Again, the German example is graphic. Each of its 25 active war-mobilized army corps consisted of 41,000 men, 14,000 horses, and 2,400 supply wagons; each corps occupied 30 miles of road upon formation; and each was to be formed at the border in 24 hours. Thereafter, each corps had to be directed during the advance, fed, and reinforced. In August 1914 the General Staff employed 11,000 trains to transport 2.1 million combatants to the front in 312 hours. More than 2,150 trains of 54 cars each rattled across the Hohenzollern Bridge at Cologne alone in ten-minute intervals between 2 and 18 August.[20]

Nor was the Prussian-German example of 1914 an exception in terms of scope. Of the Russian army's total of almost 2 million combatants organized into 114.5 infantry divisions, 52 divisions had been mobilized by the twenty-third day of mobilization (M + 23) and shuttled to the empire's western frontier by 350 trains per day. France mobilized not only the 800,000 men of its peacetime home army, but also about 1.8 million reservists for an active force of 2,689,000 men. By 18 August, fourteen rail lines, each passing 56 trains per day on average, transported this force to the front in 4,278 trains.[21] And even in "moribund" Austria-Hungary, the army railway staff of more than 47,000 officers and men was tasked with transporting 3.4 million men to the various fronts in 11,000 trains, each of 50 cars.[22]

The division of Europe by 1907 into two nearly equal camps – the Triple Alliance of Austria-Hungary, Germany, and Italy, on the one hand, and the Triple Entente of Britain, France, and Russia, on the other – added an additional complication as mobilization schedules needed to be coordinated with allies. These not only had their individual and separate military cultures, but also divergent vital interests and national ambitions. Thus, a host of staff discussions, both formal and informal, sought to revise and (it was hoped) to coordinate existing war plans.

Finally, it should be pointed out that the six analyses of war plans and war planning in this volume to a degree remain at the mercy of primary sources. Most grievously, in the German case, Allied air raids on the

[20] Germany, Reichsarchiv, *Der Weltkrieg 1914 bis 1918* (14 vols., Berlin, 1925–56), vol. 1, p. 69.

[21] Hew Strachan, *The First World War*, vol. 1, *To Arms* (Oxford, 2001), pp. 206, 297.

[22] Herwig, *First World War*, pp. 55–56. In reality, the figure of 11,000 trains was reduced to 1,942 in 1914 due to an inadequate rail network.

Reichsarchiv at Potsdam in February and April 1945 destroyed most of the records of the Prussian-German General Staff. After World War I, self-appointed "patriotic censors" pressured the widow of General von Moltke to destroy the bulk of the former staff chief's personal papers.[23] Those of his successor, Erich von Falkenhayn, likewise "disappeared" in the 1920s.

Similar "patriotic censors" were also at work in Vienna. The official diary of Foreign Minister Count Leopold Berchtold was "cleansed" of any entries for the critical week of 27 June to 5 July 1914. The secret files of the intelligence division of the General Staff fall silent for a full year after the receipt of three telegrams from Sarajevo on 28 June announcing the murder of Archduke Francis Ferdinand and his wife Sophie.[24] And the secret files on Colonel Alfred Redl's treason and subsequent suicide were concealed until found by historian Günther Kronenbitter in the 1990s.

In the Italian case, historian John Gooch states that the royal family archives were removed after 1945, first to Portugal and later (it is rumored) to Switzerland. Given that the king was the final arbiter in all military matters, this leaves a black hole in any discussion of Italy's war planning.

On the Allied side for 1914, historians of the Russian military were denied complete access to the archives from the late 1920s to about 1991. Even today, historian Bruce Menning reports, the Manuscript Section of the Russian State Public Library in Moscow does not permit ready use of the papers of one of Imperial Russia's most important war planners, General M. V. Alekseev.[25] In the British case, historian Keith Neilson shows that the Cabinet, which held ultimate authority over "war planning," kept no minutes of its meetings, and hence we are dependent for information on the crucial final stages of decision making on the later memoirs of some of its members. And with regard to France, the government did not even debate whether to go to war since the German invasion for it constituted an automatic *casus belli*. Moreover, the Chamber

23 Holger H. Herwig, "Clio Deceived: Patriotic Self-Censorship in Germany after the Great War," *International Security* 12 (1987), p. 37. Some 3,000 files from the Army Research Institute for Military History did survive the Allied bombings and were returned to the German Democratic Republic in 1988. They are today at the Federal Military Archive (BA-MA) at Freiburg: RH 61 Kriegsgeschichtliche Forschungsanstalt des Heeres, Teil 1, Teil 2.

24 Herwig, *First World War*, pp. 11, 14.

25 Bruce W. Menning, "Pieces of the Puzzle: The Role of Iu. N. Danilov and M. V. Alekseev in Russian War Planning before 1914," *The International History Review* 25 (2003), pp. 775–6.

of Deputies, which had to approve any declaration of war, began its summer recess on 15 July and did not reconvene until 4 August – after the German declaration of war. Hence, we have no indication to what degree civilian and military authorities took either the military calculus of Plan XVII or their knowledge of the basic contours of the Schlieffen Plan into account in July–August 1914.

Threat Perception and Military Planning

This notwithstanding, historians have been able to piece together the broad parameters of war planning. To varying degrees, the six major powers discussed in this volume harbored strong threat perceptions. Austria-Hungary, Kronenbitter shows, considered itself to be surrounded by foes. Although Russia was the Dual Monarchy's most dangerous rival for influence in Galicia and on the Balkan Peninsula, Serbia likewise threatened the Monarchy's security (and ambitions) in the Balkans. Even two treaty allies, Italy and Romania, were seen as potential threats. Patterns of threat perception, Kronenbitter argues, lent plausibility to Vienna's fears that "the laws of history and nature" might sooner or later turn potential opponents and even allies into deadly threats.

Translated into military terms, these perceptions made war on two or even three fronts highly possible. Given the Dual Monarchy's limited military resources, the General Staff's planning focused on one opponent first before "swirling around the bulk of the army against the other one"; decisive battlefield victories alone "came to be seen as the only way out of the strategic impasse." After the Bosnian annexation crisis of 1908–09, the most likely scenario became a two-front war against Serbia and Russia. Flexibility was essential. Hence, Conrad divided his armies (48½ divisions) into three striking forces. A-Group with at least 28 infantry divisions would be concentrated against Russia (92 divisions); Minimal Group Balkan consisting of 8 divisions would be marshaled against Serbia and Montenegro (21 divisions); and B-Group, a strategic reserve consisting of two Hungarian and two Bohemian corps, would be directed either to the Balkan theater or to Galicia (in case of Russian intervention). Obviously, Habsburg forces would be outnumbered roughly two-to-one on the two fronts – provided that Italy and Romania remained loyal allies.

The certainty of numerical inferiority made Habsburg military planning highly contingent on German planning. In Berlin, also, as historian Annika Mombauer explains, threat perceptions dominated the men of

1914. After the completion of the Anglo-French-Russian *entente cordiale* in 1907, they perceived Germany to be "encircled" by a ring of hostile powers. Austria-Hungary seemed to be the only reliable ally as Rome had not supported Berlin at Algeciras in 1906 in the aftermath of the first Moroccan crisis.

The General Staff under Alfred von Schlieffen saw relief from this perception of "encirclement" by way of a bold and desperate strategy. Like Conrad, Schlieffen accepted that the next war would have to be fought on two fronts. Arguing that Russia would be slow to mobilize due to its underdeveloped railway system and haunted by the Napoleonic nightmare of spectacular but meaningless victories in the East, Schlieffen, as is well known, opted to commit seven-eights of available forces to crush France in the West by way of a grand sweeping move through the Netherlands, Belgium, northern France, and in behind Fortress Paris. After crushing French forces against the "anvil" of German units stationed in Lorraine, the general proposed to shuttle those troops east to blunt the slowly mounting Russian assault. The key to victory, Mombauer reveals, lay in speed. Germany could ill afford a protracted war against numerically superior adversaries, and hence victory in the West was projected within forty days of mobilization.

Italy, the third member of the Triple Alliance, was also beset by a nightmare of threat perceptions. The wars of the Risorgimento, Gooch suggests, "left in their wake an enmity towards Austria which was matched by a fear of and distaste for France." To be sure, Rome had sought to enhance its security on the Continent by way of an alliance with Berlin in 1882, but that accord (Triple Alliance) included a less than popular tie to Vienna – ruler of an Italian ethnic majority (97 percent) in the Trentino as well as of an ethnic minority (9 percent) in South Tyrol. It was an unhappy trade-off and one that aroused constant concerns for the viability of the tie to Rome in both Vienna and Berlin.

Italy also nurtured colonial ambitions in the Mediterranean region and these were bound to bring it into potential conflict with France and Britain. To escape that conundrum, Italian Foreign Minister Giulio Prinetti in June 1902 had signed a diplomatic accord with French Ambassador Camille Barrère placing Morocco into the French and Tripoli-Cyrenaica into the Italian sphere of influence – while at the same time Rome renewed the Triple Alliance in 1887, 1891, 1902, and 1912, and even inked a naval accord with Austria-Hungary two years before the outbreak of war in 1914.

To complicate matters further, Gooch illustrates that "the political architecture of Liberal Italy had created a government system that neither required nor encouraged coordination between army and navy or between the forces and the bearers of the civilian portfolios." Put bluntly, there existed no coordination of grand strategy. "Italy's soldiers," in Gooch's words, simply "looked to win battles – or at least not to lose them – on the assumption that this would win whatever war the politicians decided to start."

As a result, Italy's generals planned in a political and diplomatic vacuum. The 1888 Triple Alliance called for Italian forces to be sent to the Upper Rhine River to fight alongside the German army in case of war with France. In the last peacetime renewal of the Triple Alliance in 1912, Chief of the General Staff Alberto Pollio promised his German counterpart, General von Moltke, that in case of such a war, Italy would "mobilize all her forces and would take the offensive in the Alps without delay." Moreover, Pollio seriously contemplated landing troops in the south of France – perhaps at the mouth of the Rhône River – and leading Italy's Third Army either against Verdun or even into Galicia! The Italian navy, for its part, promised to interdict French troop movements from North Africa.[26] It would fall upon Italy's political leaders to address these issues in July–August 1914.

With regard to the Triple Entente, Russia also faced the problem of a future war on two fronts. In the wake of its defeat at the hands of the Japanese in 1905, Russia was left with little strategic choice. Its treasury was empty, 1 million of its soldiers were in the Far East in a state of either demobilization or redeployment, 15 of its capital warships and their crews rested in watery graves, and revolution shook the state. The war with Japan had been one of protracted battles and extended fronts. It had demonstrated not only to the two combatants but also to foreign observers from France, Germany, Great Britain, and the United States the defensive power of entrenched automatic weapons, barbed wire, machine guns, and grenades; as well as the offensive power of indirect artillery fire.[27] In places, it had degenerated into trench warfare; in the case of the Japanese Nambu brigade it had caused a staggering 90 percent casualty rate. And it had revealed fully the deficiencies of a disunified Russian command structure.

[26] Hamilton and Herwig, eds., *Origins*, p. 364.

[27] For new interpretations, see John W. Steinberg, ed., *The Russo-Japanese War in Global Perspective: World War Zero* (Leiden, 2005).

Not surprisingly, Russian planners cut their strategic commitments by reaching a series of agreements and understandings with Japan between 1907 and 1911, which were reinforced by the growing Anglo-Russian *rapprochement* before 1914. That left St. Petersburg with the conundrum of how to pursue its interests with regard to the Ottoman Empire in the South while concurrently fulfilling its alliance commitments to France in the North. Since signing a military accord with France in 1892, Russia knew that Germany would fight a two-front war, but not whether it would concentrate first against France or against Russia. In the end, Russian military planners ascribed primary importance to the European threat – that is, to Germany in the North and Austria-Hungary in the Southwest. Hence, a possible nightmare scenario would be a war with both the Ottoman Empire and the Austro-Hungarian-German empires. The Russian General Staff saw threats everywhere, friends nowhere. "Only an invasion by Martians," military historian A. A. Kersnovskii observed, "was unforeseen."[28] Unsurprisingly, conservatism and caution dominated war planning.

To defuse what by 1912 had become a bitter "paper war" over war planning between the General Staff in St. Petersburg and the military district staffs at Kiev and Warsaw, War Minister V. A. Sukhomlinov in February convened a meeting of his staff chiefs and district commanders at Moscow to establish strategic planning priorities. The meeting hammered out the basic contours of what later came to be known as Schedule 19, in Menning's words, an "extremely risky strategy in the name of opportunistic comprise" between Iu. N. Danilov's "Germany" and Alekseev's "Austria-Hungary" offensive proposals. A "mixture of fear and opportunism" as well as "altruism" defined the need to operate offensively against Germany by M + 14; "self-preservation" and "opportunism" defined the same kind of offensive operations against Austria-Hungary. Planners resolved to direct the main thrust with fifteen-plus army corps against Austria-Hungary (Variant "A"), while not concurrently rejecting an offensive with six army corps against East Prussia (Variant "G"). In short, Russia would mount "two simultaneous offensive operations along two diverging strategic axes." Tsar Nicholas II gave his blessing to this "dangerous strategic compromise" in May 1912. "Variant A" became operative in July 1914.

Russian planning, of course, was based in large measure on the policies and plans of its ally, France. That nation's primary threat perception

[28] Cited in Menning, "Pieces of the Puzzle," p. 784.

concerned the fear of another war with Germany. To this end, political leaders in Paris had sought to shore up their position by way of a military accord with Russia in 1892, followed by an *entente cordiale* with Britain in 1904. In 1911 and 1912 the French and Russian General Staffs agreed that they would without prior consultation launch a coordinated offensive in the event of war against Germany; the Russians promised offensive operations by 800,000 troops at least by the fourteenth day of mobilization (M + 14). And in 1913 France increased mandatory male conscription from two to three years to be better able to meet the perceived threat from Germany.

Joseph Joffre, appointed chief of the General Staff in July 1911, shaped the final contours of French military planning before the outbreak of war. Historian Robert Doughty argues not only that the *Plan des renseignements* was "Joffre's own," but that it was little more than "a concentration plan with operational alternatives."[29] In all his writings, Joffre exhibited an almost fanatical adherence to the "logical and sensible doctrine" of the *offensive à outrance*. "The French army," he wrote in October 1913, "accepts no law in the conduct of operations other than the offensive.... Battles are above all moral contests."[30]

Joffre chose not to share what eventually became Plan XVII with his subordinate commanders and his staff until he published the General Instruction No. 1 on 8 August 1914. Out of fear of possible "meddling of the government in military operations," the general also kept his intentions secret (as long as he could) from his political superiors. This included the Superior Council of War (Conseil supérieure de la guerre), presided over by the minister of war, and the Superior Council of National Defense (Conseil supérieure de la défense nationale), chaired by the president and including premier, minister of war, and minister of foreign affairs.

In contrast, Joffre as well as his predecessors maintained close ties to the Russians by way of annual staff talks – especially after the poor performance of Russian units in the Russo-Japanese War – between 1906 and 1913. Given their firm belief that the Germans would concentrate the bulk of their forces in the West and leave only minimal forces in the East, the two staffs agreed by July 1912 that France would mount a "vigorous and determined" offensive against Germany with 1.3 million soldiers.

[29] Robert A. Doughty, "French Strategy in 1914: Joffre's Own," *Journal of Military History* 67 (April 2003), pp. 427–54; and Doughty, *Pyrrhic Victory: French Strategy and Operations in the Great War* (Cambridge, MA, and London, 2005), pp. 17–57.

[30] Cited in ibid., p. 26.

Of course, these intimate arrangements were not "binding" on the British government.

A major stumbling block to French planning was one critical variable: Belgium. Ever since Plan XII of 1892, French military planners had expected Germany to violate that country's neutrality in case of war, but their desire to preempt such a thrust by moving into Belgium first had repeatedly foundered on the refusal of Premiers Joseph Caillaux and Raymond Poincaré (1912) to start a war in this manner. For his part, Joffre argued that the Germans would not concentrate their active and reserve formations, and thus could not mount sufficient corps for a sustained drive through eastern Belgium in the direction of Sedan and Montmédy. His staff similarly argued that a German assault into Belgium beyond the Liège-Namur line would "dangerously overextend their front and have an insufficient density for a vigorous action."[31]

Joffre submitted his initial revisions for the French plan of concentration to the Superior Council of War in April 1913; he completed its main parts in February 1914; and his staff finished all annexes in May 1914. Joffre segmented French forces into five armies: ten corps along the eastern frontier with Germany; five watching Belgium; and six centered on Verdun, to be used as a "swing" force to slide in either direction. Plan XVII, in Doughty's analysis, was a concentration plan that placed French forces at the frontier; Joffre would decide their strategic mission once there. Plan XVII offered no strategic design; it merely stated that the commander-in-chief would "deliver, with all forces assembled, an attack against the German armies."[32] In this way, Joffre hoped to maintain flexibility until he received clear indications of enemy intentions. If the Germans invaded Belgium, as Joffre expected, he planned to hurl Fifth Army northward in the direction of Florenville and Neufchâteau. A secret annex to Plan XVII stated that if the British ("*l'armée W*") joined the war at the outset, their forces would be positioned on the left of Fifth Army near Hirson. The German sweep through Belgium in August and September 1914 would come as a shock to Joffre.

And just as Russian planning depended to a large extent on the French, so French planning to some degree was linked to what its other Entente "ally" might do in a general European war. British war planning, historian Keith Neilson argues, is perhaps the most "problematic" to analyze, for it is utterly "misleading" even to speak of British "war planning" in the

[31] Cited in Doughty, "French Strategy in 1914," p. 441.
[32] Ibid., p. 442.

strict sense of the term. Britain was not formally bound by treaty to go to war in Europe. Its *entente* with France (1904) was an understanding over colonies, spheres of interest, and fishing rights; that with Russia (1907) over spheres of influence in Persia, Afghanistan, and Tibet. Britain's prewar plans thus were "contingent upon circumstances." According to Neilson, there existed no commitment, "moral" or otherwise, for it to intervene in a continental war. Any decision for foreign intervention would be made by the prime minister in consultation with his Cabinet.

In fact, the major issue of concern in British security policy revolved around the defense of India. But after the formation of a General Staff in 1905, a number of highly professional officers in the War Office did begin to turn their attention to continental affairs. An unpredictable Imperial Germany under Kaiser Wilhelm II seemed to threaten London's time-honored policy of maintaining the "balance of power" on the Continent – that is, of not letting any hegemon attain "a position of superiority which would enable it... to dominate the other Continental Powers," and therewith Britain's largest trading market. In accordance with such a perceived "continental commitment," the British and French general staffs in 1906, 1907, and 1908 worked out contingency plans to send a British Expeditionary Force (BEF) to France to deal with the German menace.

Still, the decision remained with the Government. In August 1911 Prime Minister Herbert Henry Asquith called a meeting of the Committee of Imperial Defence (CID) to determine how Britain might give "armed support to the French," if required. While the Royal Navy under Admiral Sir Arthur Wilson advocated a close blockade of the German North Sea coast combined with a series of tip-and-run raids, Chief of the Imperial General Staff Sir William Nicholson and Director of Military Operations General Sir Henry Wilson countered with a specific proposal to dispatch six infantry divisions and a cavalry division of the BEF to fight on the left wing of the French army on the Continent. But the CID declined to resolve the dichotomy in war planning between the two services.

Asquith merely took both service suggestions under advisement. All questions of policy, the prime minister assured the king, "must be reserved for the decision of the Cabinet." It remained "outside the function of military and naval officers," Neilson quotes Asquith as saying, "to prejudice such questions." Specifically, the prime minister barred his military and naval planners from making any "communication" to "foreign staffs" that could, directly or indirectly, "commit this country to military or naval intentions." In short, the meeting of the CID in August 1911 hardly "set the course for a military confrontation between Britain and Germany."[33] The July–August Crisis 1914 would bear testimony to Asquith's firm faith in the primacy of politics.

Efforts at Implementation and Outcomes

The implementation of the various war plans of the six great powers in August 1914 revealed a deep chasm between theoretical planning and wartime reality.[34] Especially in the case of those states facing major war on two fronts – Austria-Hungary, Germany, and Russia – planners had to weigh developing threats against prewar expectations. In each case, personal proclivities and emotional biases came to the fore.

Conrad von Hötzendorf undoubtedly offers the best case of a commander beset by strong emotions and historical prejudices. Given the virtual certainty that any Austro-Hungarian war in the Balkans would bring about Russian intervention, Conrad's flexible prewar response made sense. But in 1914, Conrad blithely abandoned his political and strategic calculus and gave full rein to his passions as hostilities unfolded. His top priority remained first and foremost what he called "dog" Serbia. Hence he opted for "his favorite tactical maneuver, the out-flanking and eventual envelopment" of *both* Serbia and Russia by way of "pincer-like offensive operations" in Serbia and Poland.

Kronenbitter suggests that Conrad's personal bias gave pride of place to General Oskar Potiorek's Fifth and Sixth armies (460,000 troops) and their assault on Serbia. Conrad even stripped Second Army, his strategic reserve, of four divisions to "demonstrate" against Serbia along the Save and Drina rivers.[35] Within four days, Field Marshal Radomir Putnik defeated Potiorek at Jadar. A second attack by Potiorek on 8 September suffered a similar fate. Habsburg forces in Serbia sustained 24,000 casualties, with tens of thousands more succumbing to malaria and typhoid fever. Conrad's decision "for political reasons" to "slap Serbia" had turned into a stinging strategic defeat. Moreover, Italy's and then Romania's decisions in early August to remain neutral had forced Conrad

33 Niall Ferguson, *The Pity of War* (London, 1998), p. 65.

34 The first year of the war has been analyzed by Strachan, *First World War*, vol. 1, *To Arms*; the larger conflict by Michael S. Neiberg, *Fighting the Great War: A Global History* (Cambridge, MA, and London, 2005).

35 The following from Herwig, *First World War*, pp. 87–96; and Manfried Rauchensteiner, *Der Tod des Doppeladlers. Österreich-Ungarn und der Erste Weltkrieg* (Graz, 1993), pp. 121–31.

to maintain at least a credible border defense against the two former allies.

Worse was to come in the North. Russian and Austro-Hungarian forces collided in a series of heavy engagements along a 200-mile front southwest of the Pripet Marshes. At Lemberg (Lvov), the Russians caved in Conrad's Third Army and a general retreat ensued behind the Dunajec and San rivers; 150 miles of precious territory in the Bukovina and East Galicia were abandoned. The venerable Habsburg army suffered 100,000 dead and 220,000 wounded, and lost 100,000 prisoners of war. When Conrad finally ordered his strategic reserve up from the Danube, Second Army arrived in Galicia just in time to take part in Third Army's chaotic retreat. True to form, Conrad blamed his defeats on the Germans, whom he bitingly accused of having left him "in the lurch." Only a "quick, definitive decision" in France, Moltke lectured Conrad, could save Austria-Hungary.[36]

Initially, Schlieffen's blueprint, as amended by the Younger Moltke, unfolded with unexpected spectacular results. In the East, by late August the new commanders of Eighth Army, Generals Paul von Hindenburg and Erich Ludendorff, had transformed the planned "holding action" by 13 divisions of "greenhorns and grandfathers" into defeat of Russian First Army under P. K. Rennenkampf and Second Army under A. V. Samsonov invading East Prussia. Although outnumbered 485,000 to 173,000, Eighth Army inflicted losses of 50,000 casualties and 92,000 prisoners of war on Russian Second Army at Tannenberg, and of 100,000 men on Russian First Army at the Masurian Lakes.[37] Samsonov committed suicide.

In Lorraine, in what has been termed the Battle of the Frontiers, Crown Prince Rupprecht of Bavaria's Sixth Army and General Josias von Heeringen's Seventh Army had blunted the invasion of French General Édouard de Castelnau's First Army and August Dubail's Second Army. And to their north, Fifth Army under Crown Prince Wilhelm and Fourth Army under Duke Albrecht of Württemberg had stopped a French drive through the Ardennes and thrown General Pierre Ruffey's Third Army as well as Fernand de Langle de Cary's Fourth Army back behind the Meuse River.

[36] See the diary for 11 and 17 September 1914 of Lieutenant-Colonel Karl von Kageneck, Germany's military attaché to Vienna. BA-MA, MSg 1/2515.

[37] *Der Weltkrieg 1914 bis 1918*, vol. 2, p. 230; Dennis E. Showalter, "The Eastern Front and German Military Planning, 1871–1914–Some Observations," *East European Quarterly* 15 (1981), p. 174.

In the meantime, the great *Aufmarsch* in the West stormed through the narrow defile of Aachen and Liège. After defeating two corps of Sir John French's BEF at Mons and Le Cateau, General Alexander von Kluck's First Army and General Karl von Bülow's Second Army advanced on Paris. Moltke, now seduced by the vision of a classic double envelopment of French forces, abandoned Schlieffen's plan to encircle and to batter Fortress Paris from the West and instead ordered Sixth and Seventh armies to press their assaults west of Nancy. The French armies would be crushed between two German pincers. Victory by M + 40 seemed in the offing.

The Battle of the Marne began on 5 September, and for four days 1 million men and 3,000 heavy guns on either side clashed along a front stretching from Paris to Verdun. The vast German advance in sweltering heat had taken its toll: by 9 September, Second Army was in a state of demoralization and exhaustion and First Army was threatened with encirclement by French Sixth and Fifth armies. After ordering Bülow and Kluck to withdraw to the scrubby heights above the Aisne River, Moltke, a "broken man" in the words of his staff officers, resigned on 14 September.

The efforts of Moltke's successor, Erich von Falkenhayn, thereafter to rescue what remained of Schlieffen's encirclement panacea by turning the Allied flanks first in Picardy and then in Flanders – the so-called "race to the sea" – foundered on exhaustion, miserable weather, and shell shortages. The German official history cites "bloody losses" of 18,000 officers and 800,000 noncommissioned officers and enlisted men.[38]

Mombauer tackles head-on the "mythology" surrounding the Schlieffen Plan, "the stuff of legends." She details the arguments of the postwar "Schlieffen school," of military historians who argued that the plan could have succeeded under a more capable leader. She examines a recent "heated historiographical debate" wherein its most radical champion, Terence Zuber, argues that "there was no Schlieffen plan."[39] She convincingly rejects this new "mythology" on the basis of exhaustive research in German military records long thought lost but recently received from former East German archives.

Mombauer's further special contribution is a clear definition of how Schlieffen's successor, Helmuth von Moltke (the Younger), altered the

[38] *Der Weltkrieg 1914 bis 1918*, vol. 6, pp. 426–27.

[39] Zuber, *Inventing the Schlieffen Plan*, p. 299; Annika Mombauer, "Of War Plans and War Guilt: The Debate Surrounding the Schlieffen Plan," *Journal of Strategic Studies* 28 (2005), pp. 857–85.

original blueprint in at least two significant ways. First, fearing the effects of a potential British blockade on Germany, he canceled Schlieffen's proposal to march through neutral Holland, and instead sought to secure that country as a "windpipe" through which Germany could continue to import vital foodstuffs and raw materials. This decision resulted in the famous *coup de main* on Liège in the first week of the war. Second, Moltke expected the French not to remain on the defensive at the beginning of a war (as Schlieffen had), but rather that they would go on the offensive. And that offensive likely would come in the South. As a consequence, he "diluted" the German right wing by placing Sixth and Seventh armies in the South to guard against a French drive into Alsace-Lorraine (and the vital Saar industries beyond).

As is well known, the "Moltke Plan" failed. The Battle of the Marne in early September 1914 was a strategic defeat of the first magnitude for the Germans. Schlieffen's great gamble had failed. There was no fallback plan of operations. The peacetime German army had been shattered, its aura of invincibility pierced. The much-dreaded war of attrition had become reality.

In the South, the Germans had hoped that Italy, the third member of the Triple Alliance, would transport its Third Army to the Upper Rhine River. As late as 31 July 1914, General Luigi Cadorna, who had replaced the suddenly deceased Pollio as chief of the General Staff, planned to send as many as nine of Italy's available twelve army corps to Lorraine as the so-called "left hook" on the Rhine. King Vittorio Emanuele III's decision of 3 August in favor of neutrality rendered prewar planning obsolete.

Italy's "disappearance from the military scene," Gooch argues, had "major effects" on the war. For, had Italy's armies appeared as planned on the Rhine to help encircle the French forces in Lorraine, "their contribution might have had as decisive an effect as did their absence when war actually came." Instead, Cadorna in September 1914 set about crafting war plans against Austria-Hungary. These proved no more realistic than Italy's prewar plans. What Gooch calls Cadorna's "amazing strategic vision" featured a "dash" of 45 days from Friuli across the Isonzo River to the Ljubljana Plain; and from there on to Vienna.[40] When launched in June 1915, Cardona's "amazing strategic vision" degenerated into a dozen horrendous "battles of the Isonzo," culminating precisely two years later in catastrophic defeat at Caporetto.

[40] John Gooch, *Army, State and Society in Italy, 1870–1915* (New York, 1989), p. 170.

The prewar plans of the Entente powers likewise ran up against the hard reality of modern, industrial warfare. Under Schedule 19, *Stavka*, the Russian General Staff, in August 1914 mounted "two simultaneous offensive operations" against Austria-Hungary and Germany. In the North, poor roads, uncertain intelligence, failure to encode radio traffic, and personal animosities between Generals Rennenkampf and Samsonov allowed Hindenburg and Ludendorff to isolate and to destroy First and Second armies. By 27 August 1914, two Russian army corps had been annihilated and two others badly mauled at Tannenberg. Second Army ceased to be a fighting force. After its defeat at the Masurian Lakes, First Army retreated behind the Russian border. General Ia. G. Zhilinskii, commander of the northwest front, was relieved of command.

In the South, Russian forces were much more successful. General N. Iu. Ivanov stormed Lemberg with Russian Third and Eighth armies, driving across the Dunajec River and investing the great Habsburg fortress Przemyśl with its 100,000-man garrison. Austria-Hungary sustained 320,000 casualties and lost 100,000 prisoners of war in 17 days. Thereafter, bitter battles to the eastern slopes of the Carpathian Mountains brought further horrendous losses.

Menning argues that the Russian failure in 1914 stemmed from a variety of factors: an inadequate understanding of offensive-defensive correlations in the conduct of initial operations; an inability to apply mass in East Prussia due to promises made to the French for an offensive by M + 14; and a failure of operational command and control, especially on the northwest front. "Theater-strategic failure," Menning suggests, was the result of pursuing simultaneous offensive operations along two diverging and non-mutually supporting axes, and a willingness to accept prewar exercise-driven models as accurate indications of anticipated realities. Operational and tactical failure was due to inadequate logistics to support rates of advance, especially in Galicia; inadequate heavy artillery to support assaults on hasty fortifications; lack of tactical reconnaissance; and failure to provide for deep manpower reserves to exploit success.

Russian forces, living up to prewar promises to France, had rushed into battle everywhere prematurely and in the process had bled heavily for France. They had scored complete battlefield success against neither the Germans nor the Austro-Hungarians. Schedule 19, like Conrad's flexible offensive and Moltke's amended Schlieffen Plan, had failed to produce what its drafters had promised. *Stavka's* planners, both prewar and in

August 1914, had, Menning argues, ignored an old Russian proverb: "If you chase two rabbits you will catch neither."[41]

In an ironic twist of events, the French, who under Plan XVII had hoped vaguely to hurl their assembled forces toward central Germany, now found themselves forced to absorb the brunt of the German assault. To the very outbreak of the war, Chief of Staff Joffre for political reasons had ignored numerous indications from his intelligence bureau that the Germans would come through neutral Belgium; as well, he remained wedded to the "spirit of the offensive." Accordingly, once war broke out, he sent First and Second armies on his right east into Alsace and Lorraine, and Third, Fourth, and Fifth armies on his left northeast into the Ardennes.

Doughty argues that this complex splitting of the main French forces was evidence of Joffre's "inexperience," for it violated the "fundamental precept" laid down in Plan XVII: "to deliver, with all forces assembled, an attack against the German armies." Thus, Joffre's decision by 14 August to send First and Second armies into Alsace-Lorraine in hopes of catching the Germans by surprise made Plan XVII a potential loser.[42] Both armies were soundly defeated in the Battle of the Frontiers at a cost of 200,000 casualties.

Four days later, the massive German blow through Belgium fell. Joffre, still refusing to accept that the Germans had inserted their first line of reserves into the great *Aufmarsch*, compounded his earlier error in Alsace-Lorraine by ordering Ruffey's Third Army, Langle de Cary's Fourth Army, and Charles Lanrezac's Fifth Army to advance through the Ardennes Forest on 21 August. Belatedly, he allowed Lanrezac to wheel north to meet the German threat along the Sambre River. Both offensives were repulsed with heavy losses, as was the BEF's initial advance against the Germans at Mons and Le Cateau. The "Great Retreat" ensued. By 26 August, Joffre fully realized that the main German thrust was coming through central Belgium against his left flank. He had absorbed 260,000 casualties, including 75,000 dead.[43]

Plan XVII had failed. Joffre now showed his mettle. He shuttled forces from his right wing to the region around Paris: between 27 August and

[41] Menning, "Pieces of the Puzzle," p. 798.

[42] Doughty, *Pyrrhic Victory*, pp. 74–75.

[43] Anthony Clayton, *Paths of Glory: The French Army 1914–1918* (London, 2003), pp. 30ff; Doughty, *Pyrrhic Victory*, pp. 79–81; Douglas Porch, *The March to the Marne: The French Army 1871–1914* (Cambridge, 1981).

2 September, an average of 32 trains daily conducted this massive realignment. General Michel-Joseph Maunoury's newly constituted 150,000-man Sixth Army and Fifth Army (now commanded by General Louis Franchet d'Espèrey) were to play the major role in the decisive Battle of the Marne, halting German First and Second armies in their tracks. At the cost of another 250,000 casualties, Joffre pulled victory from the jaws of defeat.

Doughty argues that the French army of 1914 was "unprepared for the combat conditions it encountered in August 1914." The inadequacies of prewar training and doctrine became obvious once the two sides closed: the concept of bayonet charges without major artillery support evaporated quickly, and that of the *offensive à outrance* proved "completely inadequate in the face of modern firepower." Having paid "little or no attention to industrial mobilization or to economic requirements" for anything but a short war, Doughty argues, the French army "proved completed unsuited for the static, deadly battlefield it faced after the Marne." Yet, Joffre recovered from his initial defeat and improvised to meet the German threat. "Rarely in history," Doughty concludes, "has concerted, dedicated planning produced such inadequate results, and even more rarely in history has an army overcome such egregious errors."

Joffre's *miracle mérite* destroyed the Schlieffen Plan and saved France from certain defeat. But it had come at a heavy price. By sending the bulk of his forces away from "the main axis of the German advance through Belgium," Joffre had almost granted Moltke the forty-day victory that Schlieffen had designed. Only his imperturbable calm and last-minute resolution – and some unexpected support from the BEF – had saved his reputation, and France.

At first, Field Marshal Sir John French's 110,000-man BEF hardly seemed the stuff of heroics.[44] Concentrated on General Lanrezac's left at Le Cateau, on 23 August the BEF ran into the full force of Kluck's First Army at Mons and was forced to retreat some 200 kilometers. Three days later, General Horace Smith-Dorrien's II Corps suffered 8,000 casualties, again at the hands of Kluck's First Army. French panicked and suggested that the BEF fall back on the Channel ports, from where it could be evacuated by the Royal Navy. Only an emergency visit from Secretary of

[44] On the BEF, see Robin Neillands, *The Old Contemptibles: The British Expeditionary Force, 1914* (London, 2004); Tim Travers, *The Killing Ground: The British Army, the Western Front and the Emergence of Modern Warfare, 1900–1918* (London, 1987); John Gooch, *The Plans of War: The General Staff and British Military Strategy c. 1900–1916* (London, 1974).

State for War Lord Horatio Herbert Kitchener and Joffre's repositioning of French Sixth Army from Amiens to the Marne River to protect the BEF's retreat steadied French's resolve.

On 3–4 September, Joffre learned from French aviators that German First and Second armies had abandoned Schlieffen's bold sweeping design behind Paris and had instead sought a decisive battle east of the capital. He at once ordered the BEF to move up to Montmirail and Maunoury's Sixth Army to the Ourcq River in order to exploit this separation of the two main German forces. On 6 September, Field Marshal French's armies crossed the Marne and slowly advanced into the German gap. Moltke, believing that the BEF seriously threatened Bülow's Second Army, ordered a general retreat behind the Marne and Aisne rivers.

The BEF's modest role in the Battle of the Marne had come about as a result of Kitchener's and Joffre's steadfastness and resolution rather than from any British war planning. "Papa" Joffre had decided where the BEF was to be deployed and had ordered it to exploit the developing gap in the German advance.

The Royal Navy eased into war almost effortlessly. There was no mad "dash" into the Baltic Sea, no landing on the German coast or Danish Jutland, no shelling of German shore installations, no "tip-and-run" operations against the German fleet or ports. Instead, Admiral Sir John Jellicoe and the newly constituted Grand Fleet (21 Dreadnoughts and 4 battle-cruisers) remained at Rosyth (later at Scapa Flow) as a classic "fleet in being." Britain at once imposed a "distant" blockade on Germany and initiated its "information war," seizing control of the seas, capturing or destroying 287 enemy steamers, hunting down the isolated German cruiser squadrons, and transporting four infantry divisions and one cavalry division across the Channel to France without loss. "Neutrals," in Neilson's words, "were co-opted or bullied" into compliance "as appropriate." By December 1914, in Winston Churchill's words, "no German ship of war remained on any of the oceans of the world."[45] It was a classic strategy of sea denial.

In fact, across the North Sea, the world's second largest navy, the German High Sea Fleet (13 Dreadnoughts and 3 battle-cruisers), remained in port. Its builder, Admiral Alfred von Tirpitz, had counseled against war in July 1914, mainly because the battle fleet would not be ready for another decade with its building program 8 battleships and 13 cruisers behind schedule. The British with their "Trafalgar" mentality, so the

[45] Winston S. Churchill, *The World Crisis: 1911–1918* (London, 1932), p. 254.

argument ran in Berlin, simply would launch an "*immediate attack ... at the moment* of the outbreak of war."[46] Accordingly, little serious thought had been given to the possibility that the enemy might institute a "distant" blockade in the English Channel and the Orkney Islands. It is instructive to note that when Tirpitz queried Admiral Friedrich von Ingenohl, chief of the High Sea Fleet, during the last peacetime fleet maneuvers, "What will you do if they [the British] now do not come?" into the Helgoland Bight, neither Ingenohl nor Tirpitz could offer a suitable alternative.[47]

Strategic paralysis, in fact, quickly beset German naval planners. They made no effort in early September to interdict the transport of the BEF or its supplies to France. They quickly divided into two competing groups – those who wished to maintain the High Sea Fleet as a "fleet-in-being" to be used as a "bargaining chip" at the peace table, and those who demanded its immediate deployment in order to justify the vast sums already spent on it. The mere 28 U-boats on hand in 1914, in Tirpitz's words, constituted but a "museum of experiments."

These chapters also take into account the issue of war mobilization, financial and material. How well prepared were the various powers for what many of their leaders feared might become a "total" war? Had they adjusted the expenditures in recent wars such as the U.S. Civil War, Bismarck's wars of unification, and the Russo-Japanese War for twentieth-century conditions? In other words, did they fully appreciate the scale of the conflict that they so readily launched or joined?

Basically, no one had any precise understanding of these matters. The German General Staff in 1913 had calculated that the costs of a general European war might range as high as 10,000 to 11,000 million mark per annum; in reality, the war cost Germany 45,700 million mark yearly.[48] At Vienna, planners estimated the first three months of mobilization at 1,850 million crowns; by late July 1914, the figure had already been revised to 2,000 million crowns just for start-up costs.[49] In Russia, the State Bank by 5 August 1914 had to issue an emergency grant of

46 Report of Captain Wilhelm Widenmann, Germany's naval attaché to London, 30 July 1914; cited in Holger H. Herwig, *"Luxury" Fleet: The Imperial German Navy 1888–1918* (London, 1980), p. 149.

47 Cited in Albert Hopman, *Das Logbuch eines deutschen Seeoffiziers* (Berlin, 1924), p. 393.

48 Lothar Burchardt, *Friedenswirtschaft und Kriegsvorsorge. Deutschlands wirtschaftliche Rüstungsbestrebungen vor 1914* (Boppard, 1968), p. 8.

49 Strachan, *First World War*, vol. 1, *To Arms*, pp. 841–2.

1,200 million rubles above the legal maximum. And in Paris, the Banque de France that same day raised the volume of outlays for the military from 2,900 million to 12,000 million francs.[50]

Nineteenth-century ammunition tables – on average, 800 shells per field gun – were also woefully inadequate to the war of 1914. By 24 September, General Joffre tersely conceded, "rear now exhausted" of shells, "impossible to continue fighting for lack of munitions within fifteen days." In Germany, General von Falkenhayn on 14 November warned commanders that stocks of shells were down to four days' fighting. In Russia, British observer Alfred Knox reported individual armies down to seven days' supply. The BEF's field guns in Flanders by then were restricted to firing nine shells per day.[51] For Austria-Hungary, shortage of transportation compounded an equally dismal situation. The scale of "total" war had taken all by surprise.[52] In time, Germany led the way out of the impasse by creating special War Raw Materials Corporations at the urging of Walther Rathenau and Wichard von Möllendorff of German General Electric (AEG). "War socialism" became the new order of the day.

Finally, intelligence (both assessments and transmissions thereof) did not play a decisive role either in the drafting of war plans or during the opening stages of the war.[53] The authors of this volume are united in the view that planners in the major capitals devised their deployment plans relatively independent of their intelligence staffs, and at times in opposition to what those sources revealed. Intelligence that supported deployment plans was used to buttress them; intelligence that "failed to conform" to those plans was ignored.[54] Although the French and the Russians had acquired a great deal of first-rate information on the actual German and Austro-Hungarian war plans, they made little use of that information. Planners in Berlin and Vienna, in contrast, had no such

[50] Ibid., pp. 845, 849.

[51] Ibid., pp. 993–4.

[52] See Roger Chickering and Stig Förster, eds., *Great War, Total War: Combat and Mobilization on the Western Front, 1914–1918* (Cambridge, 2000).

[53] For three of the major powers, see John Ferris, ed., *The British Army and Signals Intelligence during the First World War* (Phoenix Mill, 1992); Douglas Porch, *The French Secret Service: From the Dreyfus Affair to the Gulf War* (New York, 1995); and Stefan Kaufmann, *Kommunikationstechnik und Kriegführung 1815–1944. Stufen telemedialer Rüstung* (Munich, 1996).

[54] See the damning verdict of Joseph Joffre in Christopher Andrew, "France and the German Menace," in Ernst R. May, ed., *Knowing One's Enemies: Intelligence Assessment before the two World Wars* (Princeton, 1984), p. 145.

knowledge of their enemies' deployment plans and made their own on the basis of best guesses and "suspicions."

Once the war began, all armies, and not just the Russian, sent crucial messages in clear because ciphers were cumbersome to use and speedy transmission was required. No army in 1914 had come to terms with what historian John Ferris calls "the cryptological consequences of the radio age."[55] Only one, the Austro-Hungarian, even possessed a signals intelligence service.[56] As a result, all relied on the telegraph, telephone, and dispatch riders.

Commonalities and Differences

The authors of the preceding case studies addressed a set of problems: the perceived threats in each capital; the cast of planners and their outlooks; the basic character of the various plans; the efforts at implementation of the plans in August 1914; and their outcomes in the opening weeks and months of the Great War. Each author argued for the uniqueness of their case, for the distinctness of the problems faced and the solutions attempted by strategic planners. Each presented their findings on the basis of the current state of availability of primary sources and scholarly interpretations. Each brought to the project their individual research expertise and findings.

It remains to comment briefly on the nature of those findings, on the commonalities as well as the differences of their cases. The object of the editors is *not* to be prescriptive, that is, to offer the proverbial "lessons of history." For, as George Ball, under secretary of state to Presidents John F. Kennedy and Lyndon B. Johnson, has eloquently noted with regard to another war:

> The most grievous offense will be the academicians' effort to offload the sins of this melancholy time on the military, who, skilled more with the sword than the pen, cannot adequately defend themselves against egghead *francs-tireurs* blowing beanshooters from the sanctuary of their ivory towers.[57]

Rather, we wish to point out the difficulties of strategic decision making on the basis of solid archival research into past experiences.

55 Ferris, ed., *British Army*, pp. 4–5; Porch, *French Secret Service*, pp. 42, 56ff.

56 See Rudolf Kiszling, *Die Hohe Führung der Heere Habsburg im Ersten Weltkrieg* (Vienna, 1977); Kronenbitter, "*Krieg im Frieden.*"

57 Cited in Richard K. Betts, *Soldiers, Statesmen, and Cold War Crises* (Cambridge, MA, and London, 1977), p. 3.

Major powers go to war (or ought to go to war) to achieve national political objectives – often expressed as "war aims." Thus, Frederick the Great battled from 1740 to 1763 to establish Prussia as a dominant power alongside Austria in the Holy Roman Empire of the Germanic Nation. George Washington fought to gain independence for the Thirteen Colonies from the British Crown. Abraham Lincoln carried out a Civil War to maintain the Union. Otto von Bismarck undertook three wars to unify Germany under Prussian authority. When those objectives had been attained, they restored the peace.

The case for 1914 is not as clear. We argued in *The Origins of World War I* (2003) that the major powers of Europe entered what became World War I mainly out of a sense of fear, of potent threat perceptions, and to maintain their great-power status rather than to destroy the European order. We detected in none of the capitals a desire in July 1914 massively to reorder Europe.

Once war commenced in August 1914, there emerged some general contours of what today is called "war termination," some general demands to be imposed after victory on the battlefield. As historian David Stevenson has argued, "War aims and strategy were interconnected."[58] The demands, he notes, revolved around territorial concessions, indemnities, and disarmament. Some demands, such as France's insistence that Alsace-Lorraine be returned, were absolute, but most were vague and negotiable. Austria-Hungary desired that Serbia be partitioned, and, later, that Russia lose some of its portion of Poland. After Italy entered the war in 1915, there was also discussion in Vienna that Italy cede some minor lands in the Alps. Germany aspired to control the Belgian and/or French Channel ports, and to annex the iron-producing regions of Longwy-Briey as well as parts of Russian Poland. Its better-known shopping list of vast war aims was not promulgated by Chancellor Theobald von Bethmann Hollweg until 9 September – that is, after the Battle of the Marne. Russia initially countered with the demand that it gain control over Polish foreign policy, public finances, and armed forces. As in the German case, a more extensive shopping list (Foreign Minister S. D. Sazonov's "Thirteen Points") only came in mid-September, after the setbacks in East Prussia and the victories in Galicia. It included taking parts of East Prussia and the lower Niemen River from Germany, and Galicia from Austria-Hungary.

[58] David Stevenson, *Cataclysm: The First World War as Political Tragedy* (New York, 2004), pp. 103–121.

Britain, with its global empire, developed no program for the postwar settlement in Europe. Its leaders generally agreed that a defeated Germany should lose its colonies and fleet. France, apart from the firm demand for Alsace-Lorraine, likewise enunciated no national program at first. It sought simply to survive the German onslaught. Only in December 1914 did Premier René Viviani call for the restoration of Belgian independence, "indemnities" for France's devastated northern areas, and an end to "Prussian militarism." Italy, having opted for neutrality in August 1914, would take a full year to shape its postwar demands. The only firm agreement came in the Pact of London on 4 September 1914, wherein Britain, France, and Russia agreed not to conclude a separate peace with Germany.

In short, none of the major powers in July–August 1914 went to war with clearly defined national programs of war aims or war termination. These would be defined and refined only as the butcher's bill began to mount. And they would grow in intensity as that bill escalated, especially after the bloody engagements at Verdun and the Somme in 1916.

Clausewitz also argued that national strategies should be coordinated – between allies, between civilian and military leaders and agencies, and between the national services. Again, 1914 offers ample evidence of contrary practice. Apart from France and Russia, which alone had a firm military alliance (1892) and which alone annually conducted staff talks in the years prior to 1914 to hammer out precise attack schedules in case of war, the other powers refused to coordinate their strategies. Austria-Hungary fought "its" war, first against Serbia and then belatedly against Russia. Germany fought "its" war against France, all the while expecting (or hoping) that the ally would guard East Prussia against Russian attack. Neither Berlin nor Vienna established a unified command to conduct the war. Neither drafted plans to allocate scarce war resources. And neither developed an exit strategy. Again, the conclusions are glaring and need not be labored.

Critically, none of the major powers managed to develop military-civilian coordination of strategies and policies. Instead, most kept their particular services' "war plans" in watertight compartments. The German General Staff did not share its one war plan with the government. Although Chancellor Bernhard von Bülow (1900–10) was generally aware that any future war would begin with breaking Belgian neutrality, his successor Bethmann Hollweg formally learned of the political implications of the Schlieffen Plan only in 1912, and then almost by accident. Neither chancellor had any knowledge of the plan's military calculus.

The General Staff did not brief its Austro-Hungarian ally on that plan; did not release the plan to the Imperial German Navy; and did nothing to coordinate war or food production for what many feared would be a protracted war.

In Vienna, Kaiser Franz Joseph never formally signed off on any of Conrad's contingency plans, and the Common Ministerial Council (representing the Austrian and Hungarian governments) did not include the chief of the General Staff – who could, on occasion, be invited, but only as a "guest." With the outbreak of war, Conrad von Hötzendorf moved his headquarters to Teschen, far removed from Vienna and its highly centralized bureaucracy.

In St. Petersburg, the tsar's Council of Ministers under Premier I. I. Goremykin made the decision to mobilize. The Council then, through Foreign Minister Sazonov, transmitted the decision to War Minister Sukhomlinov and Chief of Staff General N. N. Ianushkevich – with orders to smash his telephone (which he apparently did) to forestall any second thoughts by Nicholas II. Neither the Imperial Russian Navy nor the Parliament (Duma) was included by the small circle of decision makers.

In Italy, both the army and the navy were kept totally out of the decision-making process by a Cabinet of aristocrats that viewed the military with a combination of "disdain and disgust." Unsurprisingly, the decision for war was made by Prime Minister Antonio Salandra and Foreign Minister Antonio di San Giuliano. The General Staff was denied knowledge of Italy's diplomatic dealings or national policy. Only thus could its chief, General Cadorna, seriously prepare to send Third Army to the Upper Rhine at the very moment that the Government and king were deciding on neutrality.

Two powers did possess machinery that, at least at first glance, could (and should) have coordinated national strategy. In France, the Conseil supérieure de la défense nationale had been created precisely for the purpose of coordinating the nation's defense in economic, military, naval, political, and social terms. On it sat the leaders of both Government and the Services. And yet, the Conseil played no prominent role in war planning or war deployment in 1914. As a result, France went to war with the same fractured national command system as the other powers. The Foreign Office refused to share diplomatic information with the General Staff. At a tell-tale meeting of the Superior Council of War in October 1911, Joffre requested that he be briefed on national policy. When President Armand Fallières supported that request, Premier Joseph Caillaux literally told him to "shut up."[59] Obviously, there was no great desire by either generals or politicians at Paris to openly debate the strategic and political aspects of entering Belgium at the start of a war. Joffre, for his part, declined to share Plan XVII with the diplomats at the Quai d'Orsay. His operations bureau refused to accept repeated warnings from his intelligence bureau that the Germans planned to invade through central Belgium. In the end, President Poincaré headed off a formal decision – and much less a discussion in the Chamber of Deputies – on war by simply decreeing that the German invasion constituted the *casus belli*.

Britain since 1902 also had an institution capable of coordinating the national defense – the Committee of Imperial Defence. Its members came from the major ministries and the two armed services. Its job was to hammer out national security policies. But, as Neilson has shown with regard to the desultory meeting of the CID on 23 August 1911, no such coordination of strategic planning took place. The CID evaded the basic issue of *whether* Britain would take part in a continental war, and instead debated the details of what sort of commitment it ought to be *if* it came in at all.[60]

As the French and British cases clearly show, the mere existence of machinery to coordinate national strategies was insufficient.[61] It required players willing and able to weigh individual ministry and service policies and plans, to assess these against national resources of manpower and material, to appreciate enemy strategies and potentials, and then, and only then, to hammer out a common, united national strategy. Above all, it required flexible minds willing to make flexible responses. As historian Paul Kennedy argued in *The War Plans of the Great Powers*, in 1914 there existed no defensive strategies "because they were not wanted"; no strategic alternatives "because inflexibility was as much in the mind as in the railway timetables"; no fallback positions for stalemate or compromise "because a swift and absolute victory was what was wanted"; and little civilian control over the military "because very often they both had the same objectives and shared a common ideology."[62]

59 See p. 14 above.
60 Paul M. Kennedy, *War Plans of the Great Powers, 1880–1914* (London, 1979), p. 14.
61 Hew Strachan, "The Lost Meaning of Strategy," *Survival* 47 (Autumn 2005), p. 37. "During the First World War, the machineries for the integration of policy and strategy either did not exist, as in Germany, Austria-Hungary and Russia, or emerged in fits and starts, as in Britain and France."
62 Kennedy, *War Plans*, p. 19.

Coordinated national strategies, Clausewitz warned nearly two centuries ago, are extremely difficult to forge and to enact. Two scholars of World War II, one British and the other German, saw a great need after the tragedy of two world wars in the first half of the twentieth century to define what loosely may be termed "grand strategy." Michael Howard put it thus:

> Grand strategy . . . consisted basically in the mobilization and deployment of national resources of wealth, manpower and industrial capacity, together with the enlistment of those of allied and, when feasible, of neutral powers, for the purpose of achieving the goals of national policy in wartime.[63]

The German military historian Andreas Hillgruber offered a broader, if somewhat less elegant, definition of such a strategy-making process:

> Strategy [consists] of the integration by a state's leading elite of domestic and foreign policy, of military and psychological war planning and war conduct, of the economy and war industries, in order to arrive at an overarching ideological-power political concept.[64]

We conclude, on the basis of the six case studies offered here, that the great powers in 1914 failed to measure up to those standards.

[63] Cited in Strachan, "Lost Meaning of Strategy," p. 40. Howard's original contribution is in J. R. M. Butler, ed., *Grand Strategy* (6 vols., London, 1956–76), vol. 4, p. 1.

[64] Andreas Hillgruber, "Der Faktor Amerika in Hitlers Strategie 1938–1941," *Aus Politik und Zeitgschichte. Beilage zur Wochenzeitung 'Das Parlament'* 19 (11 May 1966), p. 3.

Appendix

Suggested Reading

On Wars: General

Archer, Christon, John Ferris, Holger H. Herwig, and T. H. E. Travers. *World History of Warfare*. Lincoln and London: University of Nebraska Press, 2002.

Black, Jeremy. *Why Wars Happen*. New York: New York University Press, 1998.

Blainey, Geoffrey. *The Causes of War*. New York: Free Press, 1988.

Kagan, Donald. *On the Origins of Wars and the Preservation of Peace*. New York: Doubleday, 1995.

Rotberg, Robert I., and Theodore K. Rabb, eds. *The Origin and Prevention of Major Wars*. Cambridge, MA, and London: Harvard University Press, 1988.

Stoessinger, John G. *Why Nations Go to War*. New York: St. Martin's Press, 1993.

On World War I: General

Boemeke, Manfred F., Roger Chickering, and Stig Förster, eds. *Anticipating Total War: The German and American Experiences, 1871–1914*. Cambridge and New York: Cambridge University Press, 1999.

Chickering, Roger, and Stig Förster, eds. *Great War, Total War: Combat and Mobilization on the Western Front, 1914–1918*. Cambridge and New York: Cambridge University Press, 2000.

Ferguson, Niall. *The Pity of War*. New York: Basic Books, 1999.

Goemans, Hein E. *War and Punishment: The Causes of War Termination and the First World War*. Princeton: Princeton University Press, 2000.

Hamilton, Richard F., and Holger H. Herwig, eds. *The Origins of World War I*. Cambridge and New York: Cambridge University Press, 2003.

Herrmann, David G. *The Arming of Europe and the Making of the First World War*. Princeton: Princeton University Press, 1996.

Horne, John, ed. *State, Society, and Mobilization in Europe during the First World War*. Cambridge and New York: Cambridge University Press, 1997.

Hunt, Barry, and Adrian Preston, eds. *War Aims and Strategic Policy in the Great War 1914–1918*. London: Croom Helm, 1977.
Joll, James. *1914:* The *Unspoken Assumptions*. London: Weidenfeld & Nicolson, 1968.
Kennedy, Paul M. *The War Plans of the Great Powers, 1880–1914*. London and Boston: Allen & Unwin, 1979.
Maurer, John Henry. *The Outbreak of the First World War: Strategic Planning, Crisis Decision Making and Deterrence Failure*. Westport, CT: Praeger, 1995.
May, Ernest R., ed. *Knowing One's Enemies: Intelligence Assessment before the Two World Wars*. Princeton: Princeton University Press, 1984.
Millett, Allan R., and Williamson Murray, eds. *Military Effectiveness*. Vol. 1: *The First World War*. Boston: Unwin Hyman, 1988.
Stevenson, David. *Armaments and the Coming of War, Europe 1904–1914*. Oxford: Clarendon, 1996.
Strachan, Hew. *The First World War*. Vol. 1: *To Arms*. Oxford and New York: Oxford University Press, 2001.
Westwood, J. N. *Railways at War*. San Diego: Howell-North, 1980.
Wilson, Keith, ed. *Decisions for War, 1914*. London: UCL Press, 1995.

Intelligence: Major Powers

Fergusson, Thomas G. *British Military Intelligence, 1870–1914: The Development of a Modern Intelligence Organization*. Frederick, MD: University Publications of America, 1984.
Ferris, John, ed. *The British Army and Signals Intelligence during the First World War*. Phoenix Mill: Alan Sutton, 1992.
Kaufmann, Stefan. *Kommunikationstechnik und Kriegführung 1815–1944. Stufen telemedialer Rüstung*. Munich: Fink, 1996.
Porch, Douglas. *The French Secret Service: From the Dreyfus Affair to the Gulf War*. New York: Farrar, Straus, and Giroux, 1995.

Strategy: General

Gray, Colin S. *Strategic Studies: A Critical Assessment*. Westport CT: Greenwood Press, 1982.
———. *Modern Strategy*. Oxford: Oxford University Press, 1999.
Green, Donald P., and Ian Shapiro. *Pathologies of Rational Choice Theory*. New Haven and London: Yale University Press, 1994.
Handel, Michael I. *Masters of War: Classical Strategic Thought*. London and Portland, OR: Frank Cass, 2001.
Kennedy, Paul M. ed. *Grand Strategies in War and Peace*. New Haven and London: Yale University Press 1991.
Lamborn, Alan. *The Price of Power: Risk and Foreign Policy in Britain, France, and Germany*. Boston: Unwin Hyman, 1991.
Miller, Steven E. *Military Strategy and the Origins of the First World War*. Princeton: Princeton University Press, 1985.

Murray, Williamson, MacGregor Knox, and Alvin H. Bernstein, eds. *The Making of Strategy: Rulers, States, and War*. Cambridge and New York: Cambridge University Press, 1994.
Paret, Peter, ed. *Makers of Modern Strategy: From Machiavelli to the Nuclear Age*. Princeton: Princeton University Press, 1986.
Posen, Barry R. *The Sources of Military Doctrine: France, Britain and Germany between the World Wars*. Ithaca: Cornell University Press, 1984.
Snyder, Jack. *The Ideology of the Offensive: Military Decision Making and the Disasters of 1914*. Ithaca: Cornell University Press, 1984.
Tractenberg, Marc. *History and Strategy*. Princeton: Princeton University Press, 1991.

Strategy: By Country Austria-Hungary

Cornwall, Mark, ed. *The Last Years of Austria-Hungary: Essays on Political and Military History, 1908–1918*. Exeter: Exeter University Press, 1990.
Kronenbitter, Günther. *"Krieg im Frieden." Die Führung der k.u.k. Armee und die Großmachtpolitik Österreich-Ungarns 1906–1914*. Munich: Oldenbourg, 2003.
Shanafelt, Gary W. *The Secret Enemy: Austria-Hungary and the German Alliance, 1914–1918*. New York: Columbia University Press, 1985.
Sondhaus, Lawrence. *Franz Conrad von Hötzendorf: Architect of the Apocalypse*. Boston: Humanities Press, 2000.
Tunstall, Graydon, Jr. *Planning for War against Russia and Serbia: Austro-Hungarian and German Military Strategies, 1871–1914*. New York: Columbia University Press, 1993.

France

Clayton, Anthony. *Paths of Glory: The French Army 1914–18*. London: Cassell, 2003.
Doughty, Robert A. *Pyrrhic Victory: French Strategy and Operations in the Great War*. Cambridge, MA, and London: The Belknap Press of Harvard University Press, 2005.
———. "French Strategy in 1914: Joffre's Own." *Journal of Military History* 67 (April 2003): 427–454.
Duroselle, Jean Baptiste. *La Grande Guerre de Français: l'incompréhensible*. Paris: Perrin, 1994.
Kiesling, Eugenia C. *Arming Against Hitler: France and the Limits of Military Planning*. Lawrence, KS: University Press of Kansas, 1996.
Luntinen, Pertti. *French Information on the Russian War Plans 1880–1914*. Helsinki: SHS, 1984.
Porch, Douglas. *The March to the Marne: The French Army 1871–1914*. Cambridge and New York: Cambridge University Press, 1981.
Ralston, David B. *The Army of the Republic: The Place of the Military in the Political Evolution of France, 1871–1914*. Cambridge, MA: M.I.T. Press, 1967.

Germany

Brose, Eric Dorn. *The Kaiser's Army, 1870–1918: Technological, Tactical, and Operational Dilemmas in Germany during the Machine Age.* Oxford and New York: Oxford University Press, 2001.

Bucholz, Arden. *Moltke, Schlieffen, and Prussian War Planning.* New York and Oxford: Berg, 1991.

Ehlert, Hans, Michael Epkenhans, and Gerhard P. Groß, eds. *Der Schlieffenplan. Analysen und Dokumente.* Paderborn: F. Schöningh, 2006.

Herwig, Holger H. *The First World War: Germany and Austria-Hungary, 1914–1918.* London and New York: Arnold, 1996.

Mombauer, Annika. *Helmut von Moltke and the Origins of the First World War.* Cambridge and New York: Cambridge University Press, 2001.

———. "The Battle of the Marne: Myths and Reality of Germany's 'Fateful Battle'." *The Historian* 68 (Winter 2006): 743–769.

Ritter, Gerhard. *The Schlieffen Plan: Critique of a Myth.* New York: Praeger, 1958.

Showalter, Dennis E. *Tannenberg: Clash of Empires.* Hamden, CT: Archon Books, 1992.

Great Britain

Dockrill, Michael, and David French, eds. *Strategy and Intelligence: British Policy during the First World War.* London and Rio Grande, OH: Hambledon Press, 1994.

French, David. *British Economic and Strategic Planning, 1905–1915.* London and Boston: G. Allen & Unwin, 1982.

———. *British Strategy & War Aims, 1914–1916.* London and Boston: Allen & Unwin, 1986.

Marder, Arthur J. *From the Dreadnought to Scapa Flow: The Royal Navy in the Fisher Era, 1904–1919.* Vol. I: *The Road to War, 1904–1914.* London: Oxford University Press, 1961.

Seligmann, Matthew S., ed. *Naval Intelligence from Germany: The Reports of the British Naval Attachés in Berlin, 1906–1914.* Aldershot: Ashgate, 2007.

Sumida, Jon Tetsuro. *In Defence of Naval Supremacy: Finance, Technology and British Naval Policy, 1889–1914.* Boston: Unwin Hyman, 1989.

Williamson, Samuel R., Jr. *The Politics of Grand Strategy: Britain and France Prepare for War, 1904–1914.* Cambridge, MA, and London: Harvard University Press, 1969.

Wilson, Keith. *The Policy of the Entente: Essays on the Determinants of British Foreign Policy 1904–1914.* Cambridge and New York: Cambridge University Press, 1985.

Italy

Bosworth, R. J. B. *Italy, the Least of the Great Powers: Italian Foreign Policy before the First World War.* London and New York: Cambridge University Press, 2005.

Gooch, John. *Army, State and Society in Italy, 1870–1915*. New York: St. Martin's Press, 1989.

Halpern, Paul. *The Mediterranean Naval Situation 1912–1914*. Cambridge, MA, and London: Harvard University Press, 1971.

Palumbo, Michael J. "German-Italian Military Relations on the Eve of World War I." *Central European History* XII (1979): 343–371.

Whittam, John. *The Politics of the Italian Army, 1861–1918*. London: Croom Helm, 1977.

Russia

Cimbala, Stephen J. "Steering Through Rapids: Russian Mobilization and World War I." *Journal of Slavic Military Studies* 9 (1996): 376–398.

Fuller, William C. *Civil-Military Conflict in Imperial Russia, 1881–1914*. Princeton: Princeton University Press, 1985.

Gatrell, Peter. *Government, Industry, and Rearmament in Russia, 1900–1914*. Cambridge and New York: Cambridge University Press, 1994.

Golovine [*sic*], N. N. *The Russian Campaign of 1914: The Beginning of the War and Operations in East Prussia*. Fort Leavenworth, KS: The Command and General Staff School Press, 1931.

Marshall, Alex. *The Russian General Staff and Asia, 1800–1917*. London: Routledge, 2006.

Menning, Bruce. *Bayonets before Bullets: The Imperial Russian Army, 1861–1914*. Bloomington: Indiana University Press, 1992.

———. "Pieces of the Puzzle: The Role of Iu. N. Danilov and M. V. Alekseev in Russian War Planning before 1914." *The International History Review* 25 (December 2003): 775–798.

Neilson, Keith. *Britain and the Last Tsar: British Policy and Russia, 1894–1917*. Oxford: Clarendon Press, 1995.

Rich, David A. *The Tsar's Colonels: Professionalism, Strategy, and Subversion in Late Imperial Russia*. Cambridge, MA, and London: Harvard University Press, 1998.

Index

Note: Page numbers in boldface indicate maps.

Lightning Source UK Ltd.
Milton Keynes UK
UKOW04f0103091215

264322UK00001B/86/P